I0482774

U.S.NRC
United States Nuclear Regulatory Commission
Protecting People and the Environment

2011–2012

INFORMATION DIGEST

2011–2012

INFORMATION DIGEST

NUREG-1350, Volume 23
Manuscript Completed: August 2011
Date Published: August 2011
U.S. Nuclear Regulatory Commission
Office of Public Affairs
Washington, DC 20555-0001
www.nrc.gov

ABSTRACT

The U.S. Nuclear Regulatory Commission (NRC) 2011–2012 Information Digest provides a summary of information about the NRC and the industry it regulates. It describes the agency's regulatory responsibilities and licensing activities and also provides general information on nuclear-related topics. It is updated annually.

The Information Digest includes NRC- and industry-related data in a quick reference format. Data include activities through 2010 or the most current data available at manuscript completion. The Web Link Index provides Web addresses for more information on major topics. The Digest also includes a tear-out reference sheet, the NRC Facts at a Glance.

The NRC reviewed information from industry and international sources but did not perform an independent verification. This edition contains adjustments to preliminary figures from the previous year. All information is final unless otherwise noted.

The NRC is the source for all photographs, graphics, and tables unless otherwise noted.

The agency welcomes comments or suggestions on the Information Digest. Please contact Ivonne Couret by mail at the Office of Public Affairs, U.S. Nuclear Regulatory Commission, Washington, DC 20555-0001 or by e-mail at OPA.Resource@nrc.gov.

Contents

NRC: AN INDEPENDENT REGULATORY AGENCY

Top: Public meeting.

Middle: The NRC Chairman and Commissioners at a briefing.

Bottom: Nuclear power plant control room.

MISSION

The U.S. Nuclear Regulatory Commission (NRC) is an independent agency created by Congress. The mission of the NRC is to license and regulate the Nation's civilian use of byproduct, source, and special nuclear materials in order to protect public health and safety, promote the common defense and security, and protect the environment.

The NRC's regulations are designed to protect both the public and workers against radiation hazards from industries that use radioactive materials.

The NRC's scope of responsibility includes regulation of commercial nuclear power plants; research, test, and training reactors; nuclear fuel cycle facilities; medical, academic, and industrial uses of radioactive materials; and the transport, storage, and disposal of radioactive materials and wastes.

In addition, the NRC licenses the import and export of radioactive materials and works to enhance nuclear safety and security throughout the world.

Values

The NRC adheres to the principles of good regulation—independence, openness, efficiency, clarity, and reliability. The agency puts these principles into practice with effective, realistic, and timely regulatory actions.

Strategic Goals

Safety: Ensure adequate protection of public health and safety and the environment.

Security: Ensure adequate protection in the secure use and management of radioactive materials.

Strategic Outcomes

- Prevent the occurrence of any nuclear reactor accidents.

- Prevent the occurrence of any inadvertent criticality events.

- Prevent the occurrence of any acute radiation exposures resulting in fatalities.

- Prevent the occurrence of any releases of radioactive materials that result in significant radiation exposures.

- Prevent the occurrence of any releases of radioactive materials that cause significant adverse environmental impacts.

- Prevent any instances where licensed radioactive materials are used domestically in a manner hostile to the United States.

Statutory Authority

The NRC was established by the Energy Reorganization Act of 1974 to oversee the commercial nuclear industry. The agency took over regulation formerly carried out by the Atomic Energy Commission and began operations on January 18, 1975. As previously noted, the NRC regulates

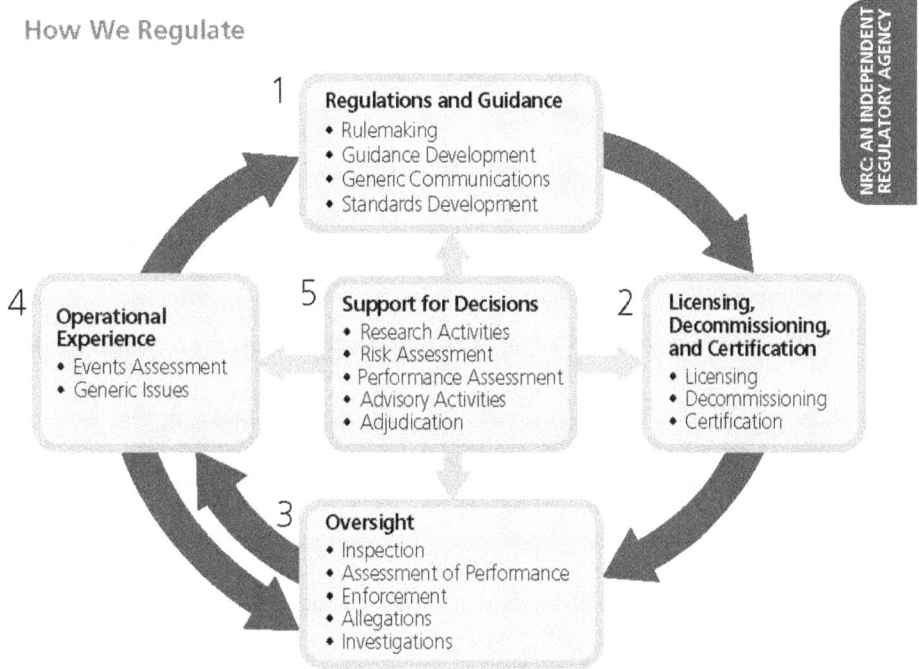

1 **Regulations and Guidance**
  - Rulemaking
  - Guidance Development
  - Generic Communications
  - Standards Development

4 **Operational Experience**
  - Events Assessment
  - Generic Issues

5 **Support for Decisions**
  - Research Activities
  - Risk Assessment
  - Performance Assessment
  - Advisory Activities
  - Adjudication

2 **Licensing, Decommissioning, and Certification**
  - Licensing
  - Decommissioning
  - Certification

3 **Oversight**
  - Inspection
  - Assessment of Performance
  - Enforcement
  - Allegations
  - Investigations

1. *Developing regulations and guidance for applicants and licensees.*
2. *Licensing or certifying applicants to use nuclear materials, operate nuclear facilities, and decommission facilities.*
3. *Inspecting and assessing licensee operations and facilities to ensure that licensees comply with NRC requirements and taking appropriate followup or enforcement actions when necessary.*
4. *Evaluating operational experience of licensed facilities and activities.*
5. *Conducting research, holding hearings, and obtaining independent reviews to support regulatory decisions.*

the civilian commercial, industrial, academic, and medical uses of nuclear materials. Effective regulation enables the Nation to use radioactive materials for beneficial civilian purposes while protecting the American people and their environment.

The NRC's regulations are contained in Title 10, *Energy,* of the *Code of Federal Regulations* (10 CFR). The following principal statutory authorities govern the NRC's work and can be found on the NRC Web site (see the Web Link Index):

- Atomic Energy Act of 1954, as amended (Pub. L. 83–703)

- Energy Reorganization Act of 1974, as amended (Pub. L. 93–438)

- Uranium Mill Tailings Radiation Control Act of 1978, as amended (Pub. L. 95–604)

- Nuclear Non-Proliferation Act of 1978 (Pub. L. 95–242)

- West Valley Demonstration Project Act of 1980 (Pub. L. 96–368)

- Nuclear Waste Policy Act of 1982, as amended (Pub. L. 97–425)
- Low-Level Radioactive Waste Policy Amendments Act of 1985 (Pub. L. 99–240)
- Diplomatic Security and Anti-Terrorism Act of 1986 (Pub. L. 107–56)
- Energy Policy Act of 1992 (Pub. L. 102–486)
- Energy Policy Act of 2005 (Pub. L. 109-58)

The NRC, licensees (those licensed by the NRC to use radioactive materials), and the Agreement States (States that assume regulatory authority over their own use of certain nuclear materials) share a common responsibility to protect public health and safety and the environment. Federal regulations and the NRC regulatory program are important elements in the protection of the public. However, because licensees are the ones using radioactive material, they bear the primary responsibility for safely handling and using these materials.

MAJOR ACTIVITIES

The NRC fulfills its responsibilities through the following licensing and regulatory activities:

- Licenses the design, construction, operation, and decommissioning of commercial nuclear power plants and other nuclear facilities, such as uranium enrichment facilities and research and test reactors.
- Licenses the possession, use, processing, handling, and importing and exporting of nuclear materials.
- Licenses the siting, design, construction, operation, and closure of low-level radioactive waste disposal sites in States under NRC jurisdiction.
- Licenses the construction, operation, and closure of a proposed geologic repository for high-level radioactive waste.
- Licenses the operators of nuclear reactors.
- Inspects licensed and certified facilities and activities.
- Certifies uranium enrichment facilities.
- Conducts light-water reactor safety research, using independent research, data, and expertise, to develop regulations and anticipate potential safety problems.
- Collects, analyzes, and disseminates information about the operational safety of commercial nuclear power reactors and certain nonreactor activities.
- Establishes safety and security policies, goals, rules, culture, regulations, and orders that govern licensed nuclear activities and interacts with other Federal agencies, including the U.S. Department of Homeland Security (DHS), on safety and security issues
- Investigates nuclear incidents and allegations concerning any matter regulated by the NRC.
- Enforces NRC regulations and the conditions of the NRC licenses, and may impose civil sanctions, including civil penalties, for violations.
- Conducts public hearings on matters of nuclear and radiological safety,

environmental concern, and common defense and security.

- Develops effective working relationships with State and Tribal governments regarding reactor operations and the regulation of nuclear materials.

- Directs the NRC program for response to incidents involving licensees and conducts a program of emergency preparedness and response for licensed nuclear facilities.

- Provides opportunities for public involvement in the regulatory process that include the following: holding open meetings, conferences, and workshops; soliciting public comments on petitions, proposed regulations and guidance documents, and draft technical reports; responding to requests for NRC documents under the Freedom of Information Act; reporting safety concerns; and providing access to thousands of NRC documents through the NRC Web site.

- Participates in Open Government initiatives that focus on open, accountable, and accessible government and engage the public in dialogue and interactions such as the use of social media and interactive high-value data sets.

The NRC hosts an annual Regulatory Information Conference attended by more than 3,000 people, including representatives from more than 30 foreign countries, the nuclear industry, and congressional staff.

ORGANIZATIONS AND FUNCTIONS

The NRC's Commission consists of five members nominated by the President and confirmed by the U.S. Senate for 5-year terms. The President designates one member to serve as Chairman, principal executive officer, and spokesperson of the Commission. The members' terms are staggered so that one Commissioner's term expires on June 30 every year. No more than three Commissioners can belong to the same political party. The members of the Commission are listed below. The Commission as a whole formulates policies and regulations governing nuclear reactor and materials

Commissioner Term Expiration

Commissioner	Expiration of Term
Gregory B. Jaczko, Chairman	June 30, 2013
Kristine L. Svinicki	June 30, 2012
George Apostolakis	June 30, 2014
William D. Magwood, IV	June 30, 2015
William C. Ostendorff	June 30, 2016

Figure 1. U.S. Nuclear Regulatory Commission Organizational Chart

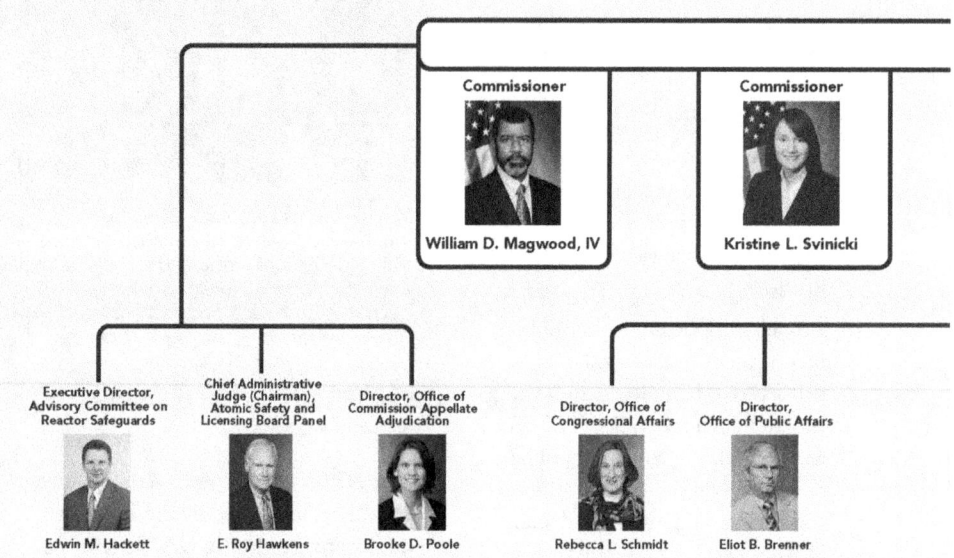

Commissioner
William D. Magwood, IV

Commissioner
Kristine L. Svinicki

Executive Director,
Advisory Committee on
Reactor Safeguards
Edwin M. Hackett

Chief Administrative
Judge (Chairman),
Atomic Safety and
Licensing Board Panel
E. Roy Hawkens

Director, Office of
Commission Appellate
Adjudication
Brooke D. Poole

Director, Office of
Congressional Affairs
Rebecca L. Schmidt

Director,
Office of Public Affairs
Eliot B. Brenner

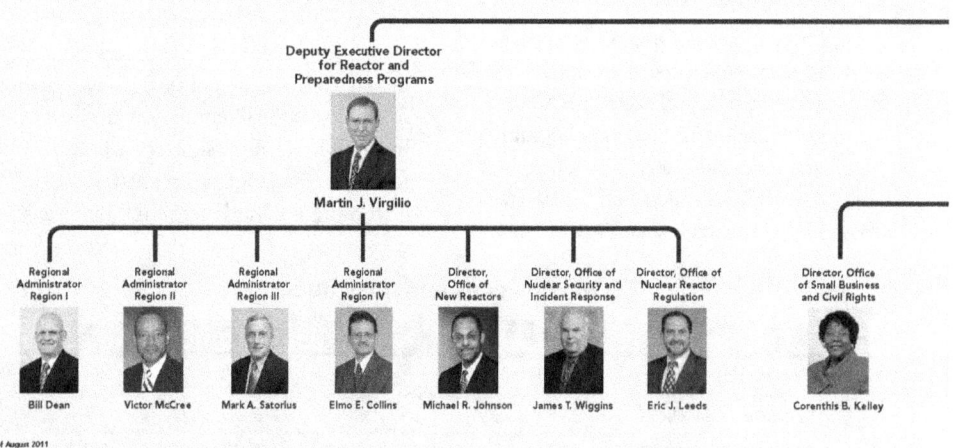

Deputy Executive Director
for Reactor and
Preparedness Programs
Martin J. Virgilio

Regional
Administrator
Region I
Bill Dean

Regional
Administrator
Region II
Victor McCree

Regional
Administrator
Region III
Mark A. Satorius

Regional
Administrator
Region IV
Elmo E. Collins

Director,
Office of
New Reactors
Michael R. Johnson

Director, Office of
Nuclear Security and
Incident Response
James T. Wiggins

Director, Office of
Nuclear Reactor
Regulation
Eric J. Leeds

Director, Office
of Small Business
and Civil Rights
Corenthis B. Kelley

As of August 2011

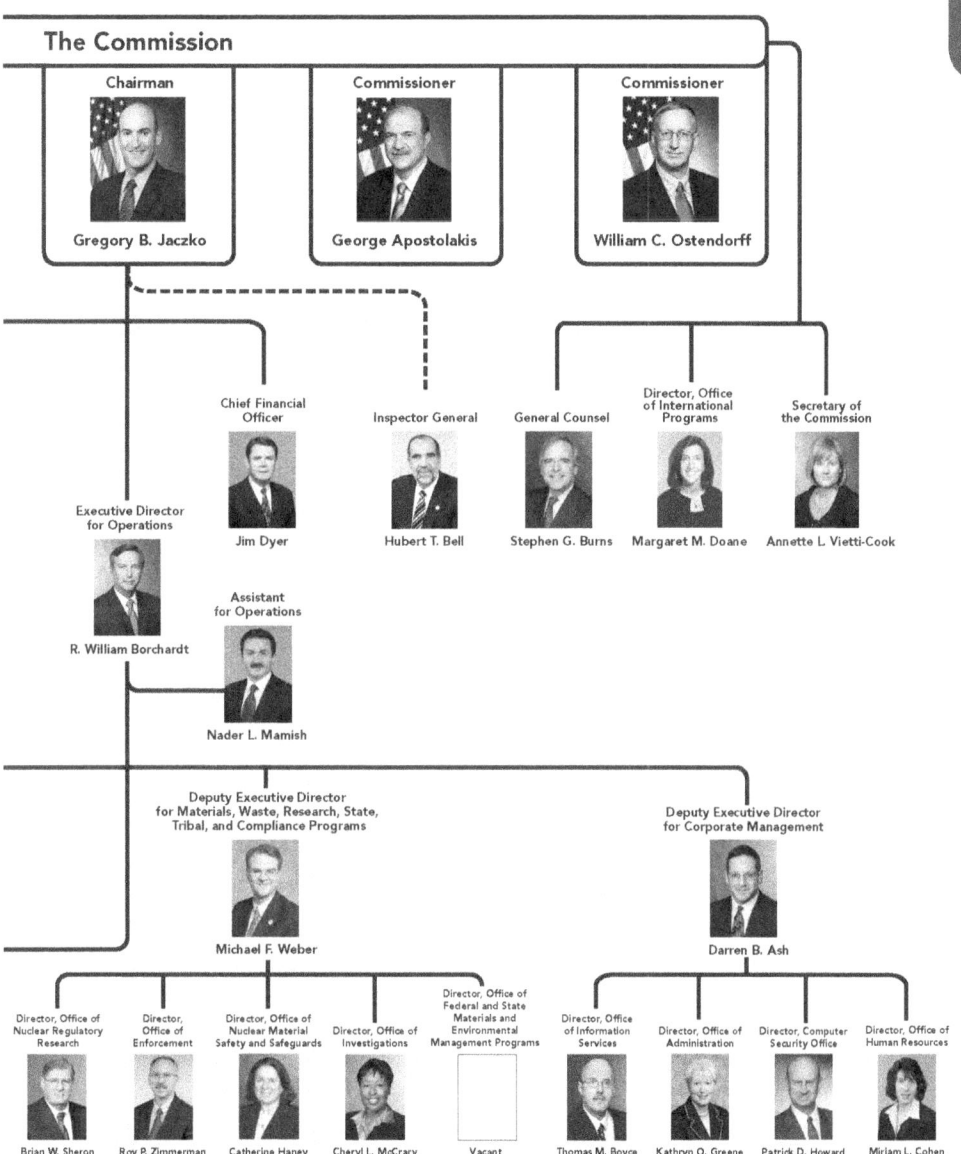

The Commission

Chairman
Gregory B. Jaczko

Commissioner
George Apostolakis

Commissioner
William C. Ostendorff

Chief Financial
Officer
Jim Dyer

Inspector General
Hubert T. Bell

General Counsel
Stephen G. Burns

Director, Office
of International
Programs
Margaret M. Doane

Secretary of
the Commission
Annette L. Vietti-Cook

Executive Director
for Operations
R. William Borchardt

Assistant
for Operations
Nader L. Mamish

Deputy Executive Director
for Materials, Waste, Research, State,
Tribal, and Compliance Programs
Michael F. Weber

Deputy Executive Director
for Corporate Management
Darren B. Ash

Director, Office of
Nuclear Regulatory
Research
Brian W. Sheron

Director,
Office of
Enforcement
Roy P. Zimmerman

Director, Office of
Nuclear Material
Safety and Safeguards
Catherine Haney

Director, Office of
Investigations
Cheryl L. McCrary

Director, Office of
Federal and State
Materials and
Environmental
Management Programs
Vacant

Director, Office
of Information
Services
Thomas M. Boyce

Director, Office of
Administration
Kathryn O. Greene

Director, Computer
Security Office
Patrick D. Howard

Director, Office of
Human Resources
Miriam L. Cohen

safety, issues orders to licensees, and adjudicates legal matters brought before it. The Executive Director for Operations carries out the policies and decisions of the Commission and directs the activities of the program and regional offices (see Figures 1 and 2).

The NRC has its headquarters in Rockville, MD, and maintains four regional offices located in King of Prussia, PA; Atlanta, GA; Lisle, IL; and Dallas, TX.

The NRC includes the major program offices described below:

- **Office of Nuclear Reactor Regulation**—handles all licensing and inspection activities associated with the operation of existing nuclear power reactors and research and test reactors.

- **Office of New Reactors**—provides safety oversight of the design, siting, licensing, and construction of new commercial nuclear power reactors.

- **Office of Nuclear Material Safety and Safeguards**—regulates activities that provide for the safe and secure production of nuclear fuel used in commercial nuclear reactors; the safe storage, transportation, and disposal of high-level radioactive waste and spent nuclear fuel; and the transportation of radioactive materials regulated under the Atomic Energy Act of 1954, as amended.

- **Office of Federal and State Materials and Environmental Management Programs**—develops and oversees the regulatory framework for the safe and secure use of nuclear materials; medical, industrial, academic, and commercial applications; uranium recovery activities; low-level radioactive waste sites; and the decommissioning of previously operating nuclear facilities and power plants. Works with Federal agencies, States, and Tribal and local governments on regulatory matters.

- **Office of Nuclear Regulatory Research**—provides independent expertise and information for making timely regulatory judgments, anticipating problems of potential safety significance, and resolving safety issues. It helps develop technical regulations and standards and collects, analyzes, and disseminates information about the operational safety of commercial nuclear power plants and certain nuclear materials activities.

- **Office of Nuclear Security and Incident Response**—oversees agency security policy for nuclear facilities and users of radioactive material. It provides a safeguards and security interface with other Federal agencies and maintains the agency emergency preparedness and incident response program.

- **Regional Offices**—conduct inspection, enforcement (in conjunction with the Office of Enforcement), investigation, licensing, and emergency response programs for nuclear reactors, fuel facilities, and materials licensees.

Figure 2. NRC Regions

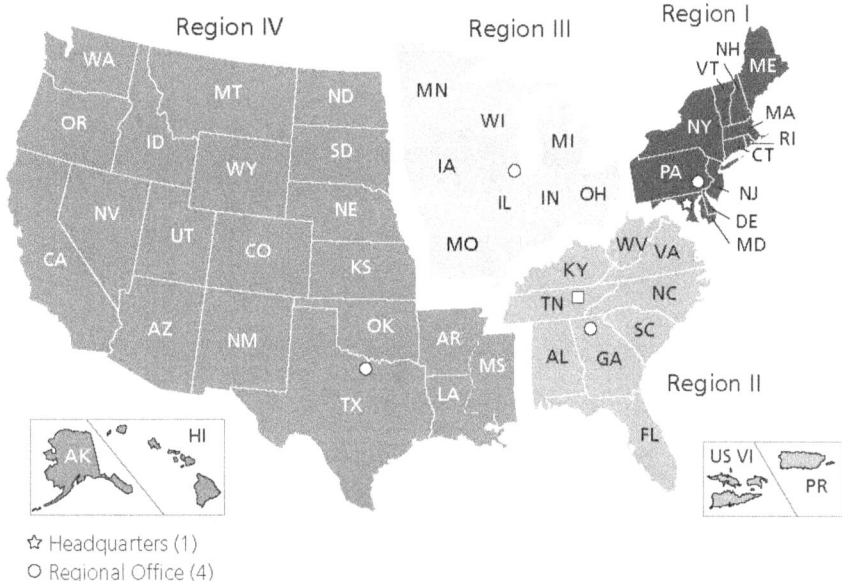

☆ Headquarters (1)
○ Regional Office (4)
□ Technical Training Center (1)

Nuclear Power Plants

- *Each regional office oversees the nuclear plants within its region, except that Region IV oversees the Grand Gulf plant in Mississippi and the Callaway plant in Missouri.*

Materials Licensees

- *Region I oversees licensees and Federal facilities located in Region I and Region II.*
- *Region III oversees licensees and Federal facilities located in Region III.*
- *Region IV oversees licensees and Federal facilities located in Region IV.*

Nuclear Fuel Processing Facilities

- *Region II oversees all the fuel processing facilities in the region and those in Illinois, New Mexico, and Washington.*
- *In addition, Region II handles all construction inspectors' activities for new nuclear power plants and fuel cycle facilities in all regions.*

BUDGET

For fiscal year (FY) 2011 (October 1, 2010–September 30, 2011), Congress appropriated $1.0541 billion to the NRC. The NRC's FY 2011 personnel ceiling is 3,992 full-time equivalent (FTE) staff

Figure 3. NRC Budget Authority, FYs 2000–2011

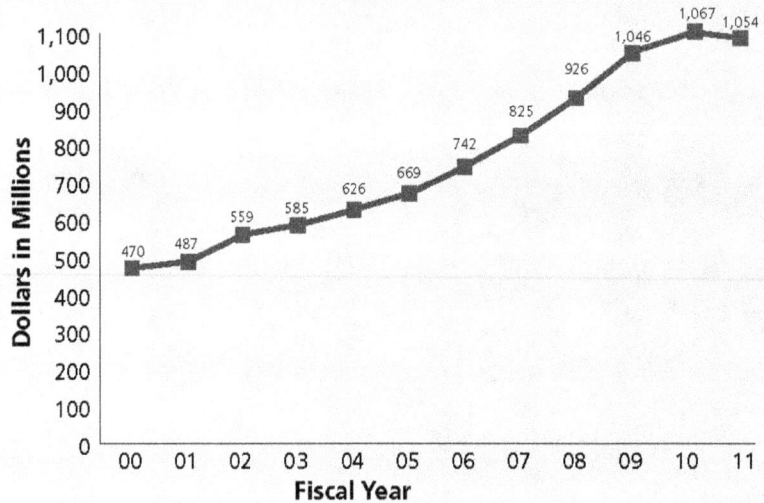

Note: Dollars are rounded to the nearest million.

Figure 4. NRC Personnel Ceiling, FYs 2000–2011

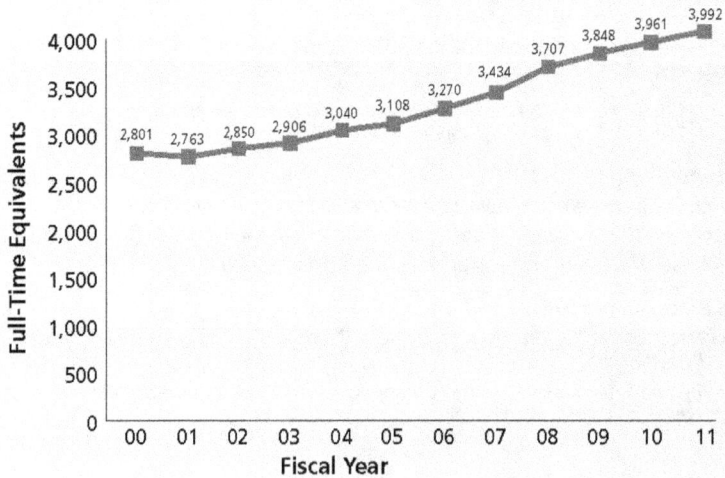

(see Figures 3, 4, and 5).

The Office of the Inspector General received its own appropriation of

$10.9 million. The amount is included in the total NRC budget. The breakdown of the budget is shown in Figure 5.

Figure 5. Distribution of NRC FY 2011 Budget Authority and Staff (Dollars in Millions)

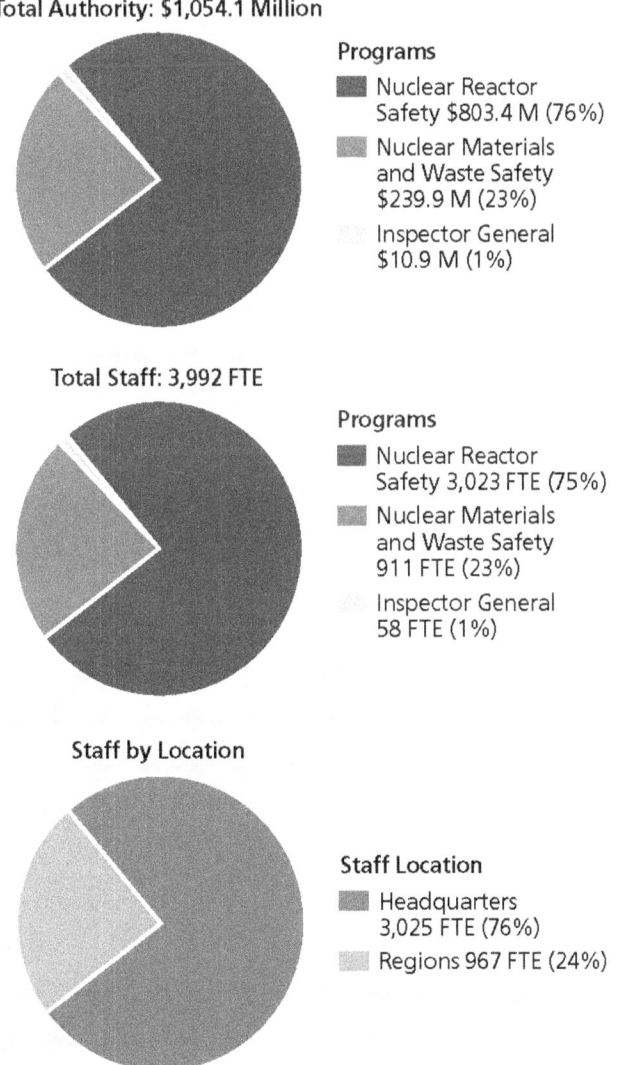

Total Authority: $1,054.1 Million

Programs
- Nuclear Reactor Safety $803.4 M (76%)
- Nuclear Materials and Waste Safety $239.9 M (23%)
- Inspector General $10.9 M (1%)

Total Staff: 3,992 FTE

Programs
- Nuclear Reactor Safety 3,023 FTE (75%)
- Nuclear Materials and Waste Safety 911 FTE (23%)
- Inspector General 58 FTE (1%)

Staff by Location

Staff Location
- Headquarters 3,025 FTE (76%)
- Regions 967 FTE (24%)

Note: Dollars and percentages are rounded to the nearest whole number.

By law, the NRC must recover, through fees billed to licensees, approximately 90 percent of its budget authority for FY 2011, less the amounts appropriated from the Nuclear Waste Fund for high-level radioactive waste activities and from general funds for waste-incidental-to-reprocessing and generic homeland security activities. The NRC collects fees each year by September 30 and transfers them to the U.S. Treasury (see Figure 6). The total budget amount to be recovered by the NRC in FY 2011 is approximately $915.8 million.

Figure 6. Recovery of NRC Budget, FY 2011*

Total Authority: $1,054 Million

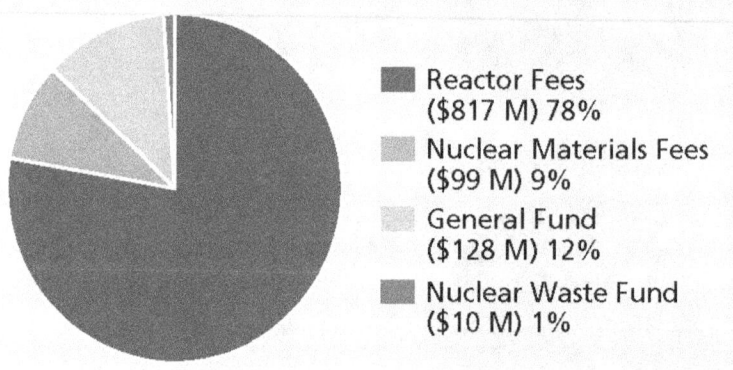

- Reactor Fees ($817 M) 78%
- Nuclear Materials Fees ($99 M) 9%
- General Fund ($128 M) 12%
- Nuclear Waste Fund ($10 M) 1%

Class of Licensee	Annual Fees
Operating Power Reactor	$4,673,000**
Fuel Facility	$589,000 to $6,085,000
Uranium Recovery Facility	$7,300 to $772,000
Materials User	$1,600 to $476,000

* Based on the final FY 2011 fee rule.

** Includes spent fuel storage/reactor decommissioning FY 2011 annual fee of $241,000.

Note. Percentages are rounded to the nearest whole number. M refers to million.

U.S. AND WORLDWIDE NUCLEAR ENERGY

Top: The NRC participates in the annual International Conference for the International Atomic Energy Agency (IAEA) in Vienna, Austria. (Photo courtesy: IAEA)

Middle: This Working Group Meeting of the Regulatory Cooperative Forum in 2010 focused on coordinating planned assistance to the Jordanian Nuclear Regulatory Commission in anticipation of Jordan's pursuit of a nuclear power program.

Bottom: Building of the Nuclear Energy Agency (NEA) in Issy-les-Moulineaux, France. (Photo courtesy: NEA)

U.S. ELECTRICITY CAPACITY AND GENERATION

U.S. electric generating capacity totaled approximately 1,025 gigawatts electric (GWe) in 2009 (see Figure 7), up slightly from 2008 (1,010 GWe). In 2009, the existing nuclear generating capacity totaled 101 GWe, which translates to 10 percent of total electric capacity.

Since the 1970s, the Nation's utilities have used power uprates as a way to generate more electricity from existing nuclear plants. By January 2011, the NRC had approved 139 power uprates, resulting in a gain of approximately 6.020 GWe at existing plants. Collectively, these uprates have added

the equivalent of six new reactors' worth of electrical generation at existing plants. Licensees responding to a December 2010 NRC survey indicated that they want to submit 35 power uprate applications in the next 5 years. If these applications are approved, the resulting uprates would add another 5.254 gigawatts (1,854 GWe) to the Nation's generating capacity (Figures 8 and 9).

As of April 2011, the 104 nuclear reactors licensed to operate accounted for 19.6 percent of U.S. net electric generation, providing 807 billion kilowatthours of electricity (see Figure 10).

Figure 7. U.S. Electric Existing Capacity by Energy Source, 2009

Total Existing Capacity: 1,025 Gigawatts Electric (GWe)

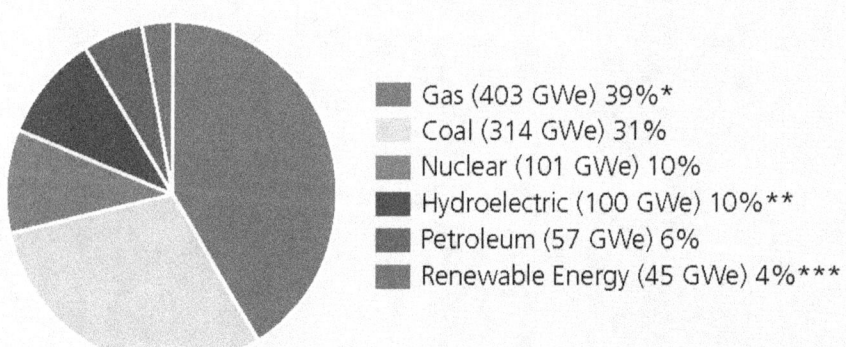

Gas (403 GWe) 39%*
Coal (314 GWe) 31%
Nuclear (101 GWe) 10%
Hydroelectric (100 GWe) 10%**
Petroleum (57 GWe) 6%
Renewable Energy (45 GWe) 4%***

* Gas includes natural gas, blast furnace gas, propane gas, and other manufactured and waste gases derived from fossil fuel.

** Hydroelectric includes conventional hydroelectric and hydroelectric pumped storage.

*** Renewable energy includes geothermal, wood and nonwood waste, wind, solar energy, and miscellaneous technologies.

Note: Totals may not equal sum of components because of independent rounding. The amounts in parentheses are measured in gigawatts electric (a gigawatt is equal to 1,000 million watts), and the information used is summer existing capacity.

Source: U.S. Department of Energy/Energy Information Administration (DOE/EIA), "Electric Power Annual," Table 1.2, "Existing Capacity by Energy Source, 2009," January 21, 2011, www.eia.doe.gov

Figure 8. Power Uprates: Past, Current, and Future

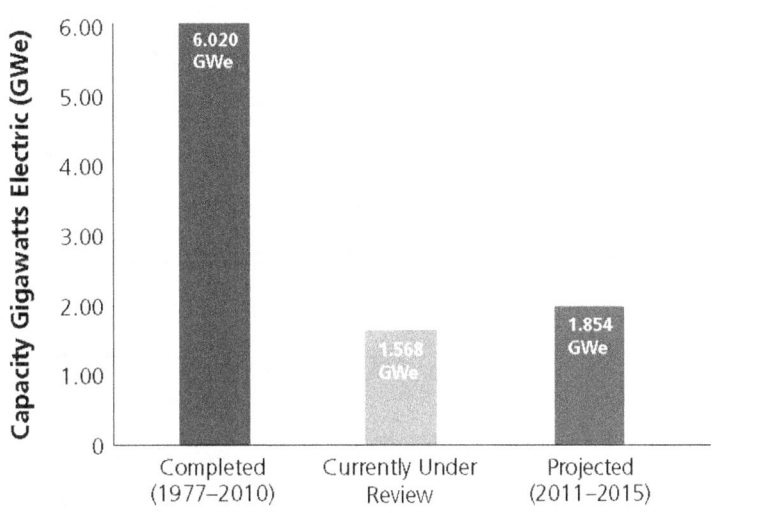

Note: Power uprates have added the equivalent of six new reactors to the U.S. power grid.
Source: December 2010 survey of NRC licensees

Figure 9. Projected Electric Capacity Dependent on License Renewals

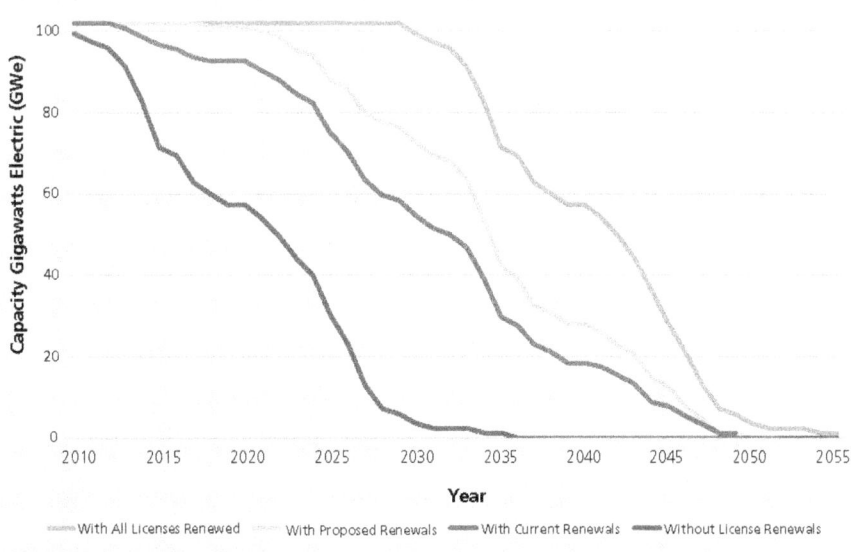

Figure 10. U.S. Net Electric Generation by Energy Source, 2000–2010

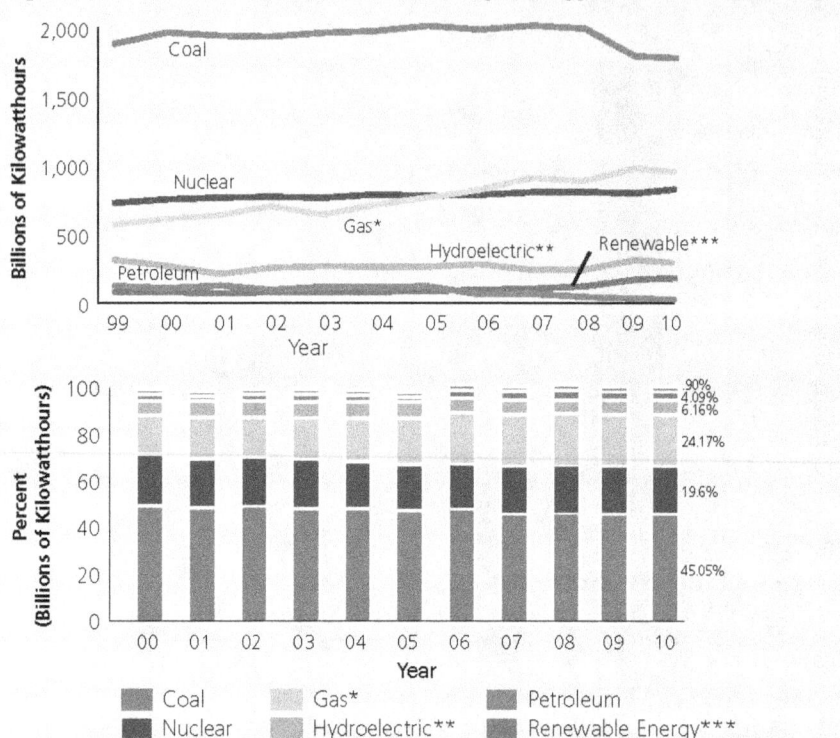

* Gas includes natural gas, blast furnace gas, propane gas, and other manufactured and waste gases derived from fossil fuel.
** Hydroelectric includes conventional hydroelectric and hydroelectric pumped storage.
*** Renewable energy includes geothermal, wood and nonwood waste, wind, and solar energy.
Source: DOE/EIA, "Monthly Energy Review," Table 7.2a, May 2011, www.eia.doe.gov

Table 1. U.S. Net Electric Generation by Energy Source, 2000–2010 (Billion Kilowatthours)

Year	Coal	Petroleum	Gas*	Hydroelectric**	Nuclear	Renewable Energy***
2000	1,966	111	614	270	754	81
2001	1,904	125	648	208	769	71
2002	1,933	95	702	256	780	79
2003	1,973	119	665	267	764	79
2004	1,977	120	726	260	788	83
2005	2,013	122	774	264	782	87
2006	1,990	64	829	283	787	96
2007	2,016	66	910	241	806	105
2008	1,986	46	895	248	806	126
2009	1,756	39	931	269	790	144
2010†	1,851	37	993	253	807	168

Note: See footnotes for Figure 10. † Based on preliminary data.
Source: DOE/EIA, "Monthly Energy Review," Table 7.2a, May 2011, www.eia.doe.gov

Table 2. U.S. Nuclear Power Reactor Average Net Capacity Factor and Net Generation, 2000–2010

Year	Number of Operating Reactors	Average Net Capacity Factor (Percent)	Net Generation of Electricity	
			Billions of Kilowatthours	Percent of Total U.S. Capacity
2000	104	88	754	19.8
2001	104	89	769	20.6
2002	104	90	780	20.2
2003	104	88	764	19.7
2004	104	90	788	19.9
2005	104	89	782	19.3
2006	104	90	787	19.4
2007	104	92	806	19.4
2008	104	91	806	19.6
2009	104	90	799	20.2
2010*	104	92	806	20.2

* Based on preliminary data.

Note: Average capacity factor is based on net maximum dependable capacity. See Glossary for definition. Refer to Appendix A for the 2005–2010 average capacity factors for each reactor. Percentages are rounded to the nearest whole number.

Source: Licensee data as compiled by the NRC

Thirty-one of the 50 States generate electicity from nuclear power plants. As of April 2009, four States (New Jersey, South Carolina, Connecticut, and Vermont) relied on nuclear power for more than 50 percent of their electricity. The data cited reflect the percentages of the total net generation in these States that were from nuclear sources. An additional 13 States relied on nuclear power for 25 to 50 percent of their electricity (see Figure 11).

Since 2000, net nuclear electric generation has increased by 6.6 percent, and coal-fired electric generation has decreased by 6.2 percent (see Table 1). Renewable energy has had the largest increase by 52 percent.

U.S. ELECTRICITY GENERATED BY COMMERCIAL NUCLEAR POWER

In 2010, net nuclear-based electric generation in the United States was a total of 806 billion kilowatthours and the average U.S. net capacity factor was 92 percent. Average U.S. net capacity factor—the ratio of electricity generated to the amount of energy that could have been generated—has increased by approximately 16 percent since 2000 (see Table 2).

Figure 11. Net Electricity Generated in Each State by Nuclear Power

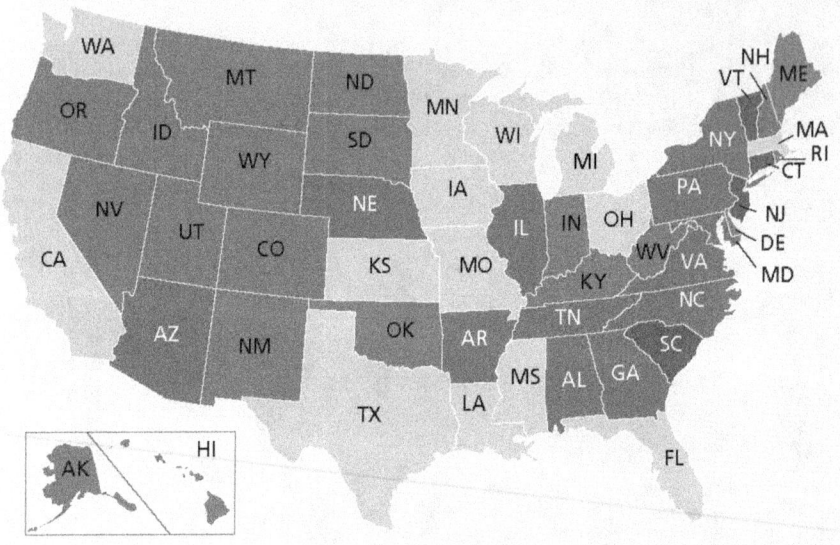

Percent Net Generation from Nuclear Sources

None (19)			1% to 24% (14)			25% to 50% (13)			More than 50% (4)		
State	Net Capacity	Net Generation	State	Net Capacity	Net Generation	State	Net Capacity	Net Generation	State	Net Capacity	Net Generation
Alaska	0	0	California	7	16	Alabama	16	28	Connecticut	26	53
Colorado	0	0	Florida	7	15	Arkansas	12	26	New Jersey	22	56
Delaware	0	0	Iowa	4	9	Arizona	15	27	South Carolina	27	52
Hawaii	0	0	Kansas	9	19	Georgia	11	25	Vermont	55	74
Idaho	0	0	Louisiana	8	17	Illinois	26	49			
Indiana	0	0	Massachusetts	5	14	Maryland	14	33			
Kentucky	0	0	Michigan	13	22	Nebraska	16	28			
Maine	0	0	Minnesota	11	24	New Hampshire	30	44			
Montana	0	0	Mississippi	8	23						
Nevada	0	0	Missouri	6	12	New York	13	33			
North Dakota	0	0	Ohio	6	11	North Carolina	18	34			
New Mexico	0	0	Texas	5	10	Pennsylvania	21	35			
Oklahoma	0	0	Washington	4	6	Tennessee	16	34			
Oregon	0	0	Wisconsin	9	21	Virginia	14	40			
Rhode Island	0	0									
South Dakota	0	0									
Utah	0	0									
West Virginia	0	0									
Wyoming	0	0									

Note: Percentages are rounded to the nearest whole number. Units measured are in megawatts.
Net summer capacity data are from the State historical tables.

Source: DOE/EIA, "State Electricity Profiles," data from April 2011, www.eia.doe.gov

WORLDWIDE ELECTRICITY GENERATED BY COMMERCIAL NUCLEAR POWER

As of June 2011, there were 440 operating reactors (at least partially) in 29 countries and Taiwan with a total installed capacity of 375,274 GWe (see Figure 12). In addition, five nuclear power plants were in long-term shutdown, and 64 nuclear power plants were under construction.

See Appendix J for the number of nuclear power reactors by nation and Appendix K for nuclear power units by reactor type, worldwide.

The United States produced approximately 27 percent of the world's gross nuclear-generated electricity in 2010 (see Figure 13). France was the next highest producer at 17 percent. Based on preliminary data in 2010, France had the highest nuclear portion (74 percent) of total domestic energy generated. In the United States, nuclear energy accounted for nearly 20 percent of the domestic energy generated (see Figure 14).

Reactors in the United States had the highest nuclear power production at 806,968 (GWe). France was the next highest producer at 407,900 (GWe) (see Table 3).

U.S. AND WORLDWIDE NUCLEAR ENERGY

Table 3. Top Ten Countries with Most Commercial Nuclear Power Reactors and Total Nuclear Power Production, 2010

Country	Number of Operating Reactors	Nuclear Power Production (GWe)
United States	104	806,968
France	59	407,900
Japan	54	279,229
Russia	32	155,107
Korea, South	21	141,894
India	19	20,480
United Kingdom	19	56,440
Canada	18	85,219
Germany	17	133,012
Ukraine	15	83,800

Note: The country's short-form name is used.

Source: IAEA, Power Reactor Information System database

Figure 12. Operating Nuclear Power Plants Worldwide

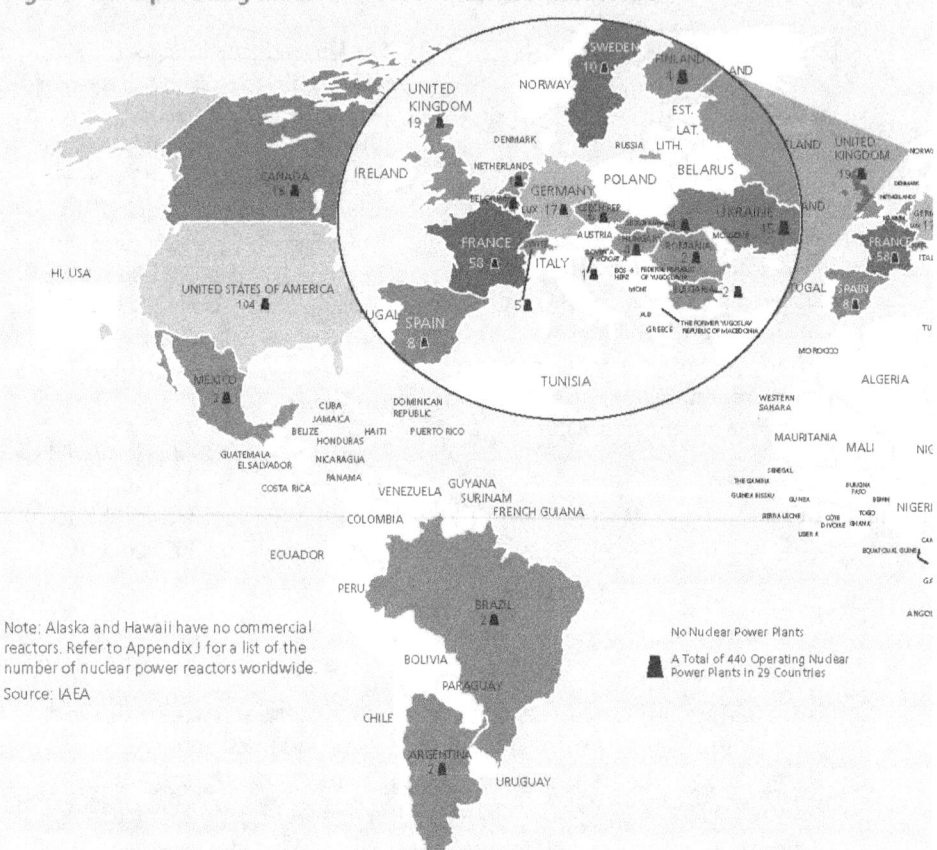

Note: Alaska and Hawaii have no commercial reactors. Refer to Appendix J for a list of the number of nuclear power reactors worldwide.

Source: IAEA

No Nuclear Power Plants

A Total of 440 Operating Nuclear Power Plants in 29 Countries

Figure 13. Gross Nuclear Electric Power as a Percent of World Nuclear Generation, 2010

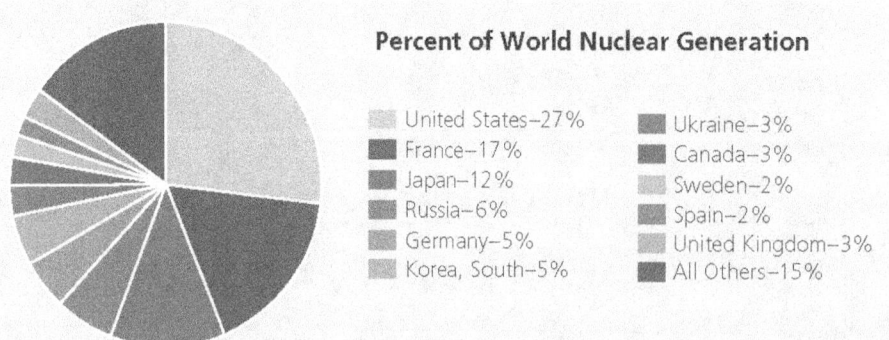

Percent of World Nuclear Generation

United States–27%	Ukraine–3%
France–17%	Canada–3%
Japan–12%	Sweden–2%
Russia–6%	Spain–2%
Germany–5%	United Kingdom–3%
Korea, South–5%	All Others–15%

Note: Because of independent rounding, the figures may not add up to the total percentage. The country's short-form name is used.

Source: IAEA, Power Reactor Information System database, as of April 26, 2011

Figure 14. Total Domestic Electricity Generation, 2010

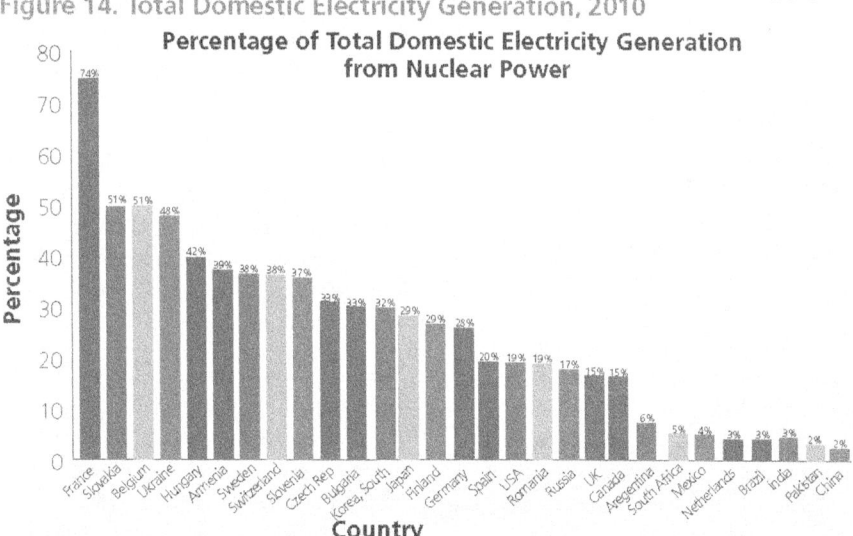

Note: The country's short-form name is used. The nuclear share in Taiwan, China is 19.3%.

Source: IAEA, Power Reactor Information System database, as of June 8, 2011

The NRC performs certain legislatively mandated international duties. These include licensing the import and export of nuclear materials and equipment and participating in activities supporting U.S. Government compliance with international treaty and agreement obligations. The NRC has bilateral programs of assistance or cooperation with 40 countries, including Taiwan, and the European Atomic Energy Community (see Table 4). The NRC has also supported U.S. Government nuclear safety initiatives with countries in Europe, Africa, Asia, and Latin America. In addition, the NRC actively cooperates with multinational organizations, such as the International Atomic Energy Agency (IAEA) and the Nuclear Energy Agency (NEA), a part of the Organisation for Economic Co-operation and Development. The NRC also has a robust international cooperative research program.

Since its inception, the agency has hosted over 340 foreign nationals in on-the-job training assignments at NRC Headquarters and the regional offices. The NRC's Foreign Assignee Program helps instill regulatory awareness, capabilities, and commitments in foreign assignees. It also helps to enhance the regulatory expertise of both foreign assignees and NRC staff. Additionally, the program improves international channels of communication through interaction with the international nuclear community and development of relationships with key personnel in foreign regulatory agencies.

Through its export/import authority, the NRC upholds the U.S. Government goals of limiting the proliferation of materials that could be used in weapons and supports the safe and secure use of civilian nuclear and radioactive materials worldwide. In addition to its direct export/import licensing role, the NRC consults with other U.S. Government agencies on international nuclear commerce activities falling under their authority. The NRC continues to work to strengthen the export/import regulations of nuclear equipment and materials, and to improve communication between domestic and international stakeholders.

The NRC assists in implementing the U.S. Government's international nuclear policies through developing legal instruments that address nuclear nonproliferation, safety, international safeguards, physical protection, emergency notification and assistance, spent fuel and waste management, and liability. The NRC also participates in the negotiation and implementation of U.S. bilateral agreements for peaceful nuclear cooperation under Section 123 of the U.S. Atomic Energy Act of 1954, as amended. The NRC also ensures licensee compliance with the U.S. Voluntary Safeguards Offer agreement and its additional protocol to the U.S.-IAEA Agreement for the Application of Safeguards in the United States.

The NRC also participates in a wide range of mutually beneficial international exchange programs that enhance the safety and security of peaceful nuclear activities worldwide. These low-cost, high-impact programs provide joint cooperative activities and assistance to other countries to develop and improve regulatory organizations. The NRC engages in the following activities:

- Cooperates with countries with mature nuclear programs to ensure the timely exchange of applicable nuclear safety and security information relating to operating reactors and consults with these countries on new reactor-related activities.

- Ensures prompt notification to foreign partners about U.S. safety issues, notifies NRC program offices about foreign safety issues, and shares security information with selected countries.

- Initiates bilateral discussions in such regulatory areas as licensing, inspection, and enforcement with countries that have recently built facilities or have vendors of equipment that may be imported to the United States during the anticipated construction of new nuclear power plants.

- Participates in the Multinational Design Evaluation Program, which leverages the resources of interested regulatory authorities to review new designs of nuclear power reactors.

Table 4. Bilateral Information Exchange and Cooperation Agreements with the U.S. Nuclear Regulatory Commission

Agreement Country	Renewal Date	Agreement Country	Renewal Date
Argentina	2012	Kazakhstan	2014
Armenia	2012	Korea, South	2015
Australia	2013	Lithuania	2010
Belgium	2014	Mexico	2012
Brazil	2014	Netherlands	2013
Bulgaria*	2011	Peru	Open-Ended
Canada	2012	Philippines	Open-Ended
China	2013	Poland	2015
Croatia	2013	Romania	2016
Czech Republic	2014	Russia*	2001
Egypt	1991	Slovakia	2015
EURATOM	2014	Slovenia	2015
Finland*	2011	South Africa	2010
France	2013	Spain	2015
Germany	2012	Sweden*	2011
Greece	2013	Switzerland	2012
Hungary	2012	Ukraine	2016
Indonesia	2013	United Arab Emirates	2015
Israel	2010	United Kingdom	2013
Italy	2015	Vietnam	2013
Japan	2015		

* In negotiation.

Note: The NRC also provides support to the American Institute in Taiwan. Egypt's agreement has been deferred until its regulatory body requests reinstatement. The country's short-form name is used.

EURATOM—The European Atomic Energy Community

The NRC hosted 20 of its international counterparts in October 2010 as part of IAEA's Integrated Regulatory Review Service, which compares a member country's nuclear regulatory approach to the international safety standards and good practices.

- Participates in a variety of conventions, treaties, and other legal and political instruments that together make up the international nuclear regime. For example, in 2010, the NRC supported the U.S. delegation to the Review Conference of the Parties to the Treaty on the Non-Proliferation of Nuclear Weapons. The NRC also participated in the preparatory Organizational Meeting for the 2011 Review Meeting of the Convention on Nuclear Safety, and in the annual meeting of countries that implement the IAEA Code of Conduct on the Safety and Security of Radioactive Sources.

- Provides guidance about export/import licensing for nuclear materials and equipment published in 10 CFR Part 110, "Export and Import of Nuclear Equipment and Material." A final rule revised 10 CFR Part 110 in July 2010 to implement updates and clarifications to the regulation. The changes became effective in August 2010.

- Assists other countries in developing and improving regulatory programs through training, workshops, peer review of regulatory documents, working group meetings, and exchanges of technical information and specialists.

- Assists countries to ensure regulatory control over radioactive sources through development of standards and provision of training and workshops through a pilot program begun in 2008. In 2010, the program was expanded to include Latin America.

- Participates in the multinational programs of IAEA, NEA, and the European Union concerned with safety research and regulatory matters, radiation protection, risk assessment, emergency preparedness, waste management, transportation, safeguards, physical protection, security, standards development, training, technical assistance, and communications.

- Participates in the International Nuclear Regulators Association meetings to influence and enhance nuclear safety from the regulatory perspective. Association members are the most senior officials of well-established independent national nuclear regulatory organizations. Current members are Canada, France, Germany, Japan, South Korea, Spain, Sweden, the United Kingdom, and the United States.

- Meets, through the NRC's Advisory Committee on Reactor Safeguards (ACRS), with other international advisory committees through annual working group meetings and plenary meetings every 4 years to exchange information.

- Participates in joint cooperative research programs through approximately 100 multilateral agreements with 30 countries and Taiwan to leverage access to foreign test facilities not otherwise available to the United States. Access to foreign test facilities expands the NRC's knowledge base and contributes to the efficient and effective use of the NRC's resources in conducting research on high-priority safety issues.

- In 2010, the NRC participated in an IAEA Integrated Regulatory Review Service (IRRS) An IRRS assesses a country's regulatory infrastructure against the international safety standards and good practices as determined by a team of international senior nuclear regulators and observers from around the world. The IRRS is carried out in three phases: a preliminary self-assessment, the peer review/mission, and a followup self-assessment and peer review.

Immediately after the March 11, 2011, earthquake and tsunami in Japan, a team of subject matter experts on reactor safety, protective measures, and international relations from the NRC and the U.S. Departments of Energy and of Health and Human Services traveled to Japan to help the Government of Japan assess and address the emergency at the Fukushima Dai-ichi nuclear power

International Nuclear Events Scale

The International Nuclear Event Scale (INES) is a worldwide tool for communicating to the public in a consistent way the safety significance of nuclear and radiological events. The scale explains the significance of events from a range of activities, including industrial and medical use of radiation sources, operations at nuclear facilities, and transport of radioactive material.

Events are classified on the scale at seven levels. Levels 1–3 are called "incidents" and Levels 4–7 "accidents." The scale is designed so that the severity of an event is about 10 times greater for each increase in level on the scale. Events without safety significance are called "deviations" and are classified as Below Scale or at Level 0.

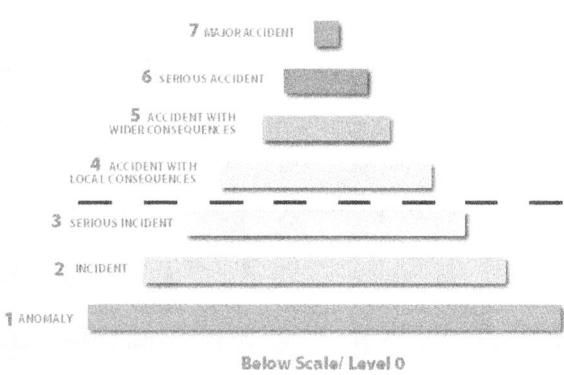

7 MAJOR ACCIDENT
6 SERIOUS ACCIDENT
5 ACCIDENT WITH WIDER CONSEQUENCES
4 ACCIDENT WITH LOCAL CONSEQUENCES
3 SERIOUS INCIDENT
2 INCIDENT
1 ANOMALY

Below Scale/ Level 0
NO SAFETY SIGNIFICANCE

plant. The NRC, which maintains
a long working relationship with its
regulatory counterpart, the Japanese
Nuclear and Industrial Safety Agency
(NISA), established a dialogue with
NISA that developed into daily
discussions about the status of the
Fukushima Dai-ichi plant's reactors
and related concerns. An NRC team
was stationed in Tokyo on March 13,
supported by additional experts working
in the NRC Headquarters Operations
Center. Approximately 30 staff
members worked in Tokyo during the
initial stages of the event on a rotating
basis in coordination with their NISA
counterparts and meeting with officials
from the Japan Nuclear Energy Safety
organizations; Tokyo Electric Power
Company; the Ministry of Economy,
Trade and Industry; the Ministry of
Education, Culture, Sports, Science
and Technology; and the Ministry of
Foreign Affairs. The NRC has reduced
its staff in Tokyo and has formed a task
force to review the events in Japan. The
task force will report findings and make
recommendations for improvements
to agency requirements, programs, and
processes. The task force recommended
short-term implementations to the
Commission in July 2011. A longer
term effort will address additional
topics related to the Japan event.

NUCLEAR REACTORS

Top: San Onofre Nuclear Generating Station, located near San Clemente, CA.
(Photo courtesy: SoCal Edison)

Middle: NRC employees lead an inspection of safety-related equipment intended for
ultimate use in nuclear facilities to ensure compliance with NRC safety requirements.

Bottom: A reactor vessel head.

U.S. COMMERCIAL NUCLEAR POWER REACTORS

As of August 2011, 104 commercial nuclear power reactors were licensed to operate in 31 States (see Figure 15). These reactors have the following characteristics:

- 4 different reactor vendors
- 26 operating companies
- 80 different designs
- 65 sites

See Appendix A for a listing of reactors and their general licensing information and Appendix L for Native American Reservations or Trust lands near nuclear power plants.

Diversity

Although there are many similarities, each reactor design can be considered unique. Figure 16 shows a typical pressurized-water reactor (PWR), and Figure 17 shows a typical boiling-water reactor (BWR). Currently there are 35 BWR and 69 PWR reactor designs.

Prairie Island Nuclear Power Plant, located near Minneapolis, MN.

Vermont Yankee Nuclear Power Plant, located near Brattleboro, VT.

South Texas Project nuclear plant, located near Bay City, TX.

Indian Point Energy Center, located near New York City, NY.

Figure 15. U.S. Operating Commercial Nuclear Power Reactors

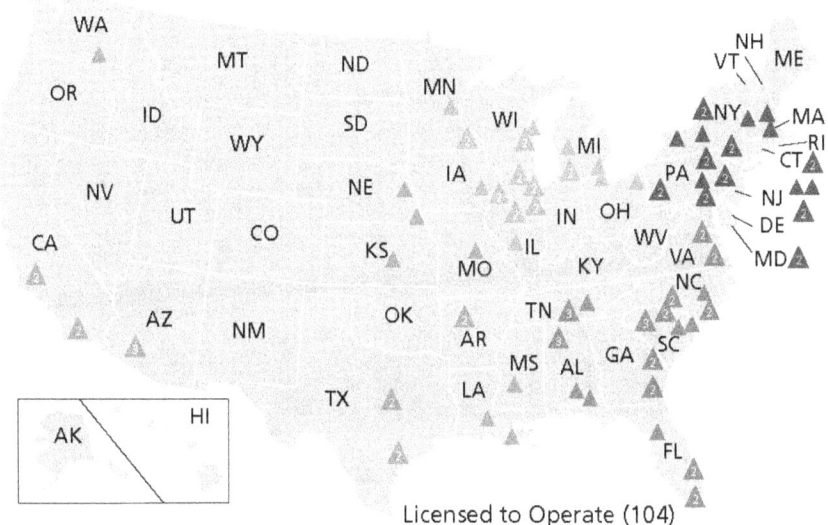

Licensed to Operate (104)

REGION I

CONNECTICUT
▲ Millstone 2 and 3

MARYLAND
▲ Calvert Cliffs 1 and 2

MASSACHUSETTS
▲ Pilgrim

NEW HAMPSHIRE
▲ Seabrook

NEW JERSEY
▲ Hope Creek
▲ Oyster Creek
▲ Salem 1 and 2

NEW YORK
▲ FitzPatrick
▲ Ginna
▲ Indian Point 2 and 3
▲ Nine Mile Point 1 and 2

PENNSYLVANIA
▲ Beaver Valley 1 and 2
▲ Limerick 1 and 2
▲ Peach Bottom 2 and 3
▲ Susquehanna 1 and 2
▲ Three Mile Island 1

VERMONT
▲ Vermont Yankee

REGION II

ALABAMA
▲ Browns Ferry 1, 2, and 3
▲ Farley 1 and 2

FLORIDA
▲ Crystal River 3
▲ St. Lucie 1 and 2
▲ Turkey Point 3 and 4

GEORGIA
▲ Edwin I. Hatch 1 and 2
▲ Vogtle 1 and 2

NORTH CAROLINA
▲ Brunswick 1 and 2
▲ McGuire 1 and 2
▲ Harris 1

SOUTH CAROLINA
▲ Catawba 1 and 2
▲ Oconee 1, 2, and 3
▲ Robinson 2
▲ Summer

TENNESSEE
▲ Sequoyah 1 and 2
▲ Watts Bar 1

VIRGINIA
▲ North Anna 1 and 2
▲ Surry 1 and 2

REGION III

ILLINOIS
▲ Braidwood 1 and 2
▲ Byron 1 and 2
▲ Clinton
▲ Dresden 2 and 3
▲ LaSalle 1 and 2
▲ Quad Cities 1 and 2

IOWA
▲ Duane Arnold

MICHIGAN
▲ Cook 1 and 2
▲ Fermi 2
▲ Palisades

MINNESOTA
▲ Monticello
▲ Prairie Island 1 and 2

OHIO
▲ Davis-Besse
▲ Perry

WISCONSIN
▲ Kewaunee
▲ Point Beach 1 and 2

REGION IV

ARKANSAS
▲ Arkansas Nuclear 1 and 2

ARIZONA
▲ Palo Verde 1, 2, and 3

CALIFORNIA
▲ Diablo Canyon 1 and 2
▲ San Onofre 2 and 3

KANSAS
▲ Wolf Creek 1

LOUISIANA
▲ River Bend 1
▲ Waterford 3

MISSISSIPPI
▲ Grand Gulf

MISSOURI
▲ Callaway

NEBRASKA
▲ Cooper
▲ Fort Calhoun

TEXAS
▲ Comanche Peak 1 and 2
▲ South Texas Project 1 and 2

WASHINGTON
▲ Columbia

Note: NRC-abbreviated reactor names listed.

▲ = 1 units ▲ = 2 units ▲ = 3 units

Figure 16. Typical Pressurized-Water Reactor

How Nuclear Reactors Work

In a typical design concept of a commercial PWR, the following process occurs:

1. *The core inside the reactor vessel creates heat.*
2. *Pressurized water in the primary coolant loop carries the heat to the steam generator.*
3. *Inside the steam generator, heat from the primary coolant loop vaporizes the water in a secondary loop, producing steam.*
4. *The steamline directs the steam to the main turbine, causing it to turn the turbine generator, which produces electricity.*

The unused steam is exhausted to the condenser, where it is condensed into water. The resulting water is pumped out of the condenser with a series of pumps, reheated, and pumped back to the steam generator. The reactor's core contains fuel assemblies that are cooled by water circulated using electrically powered pumps. These pumps and other operating systems in the plant receive their power from the electrical grid. If offsite power is lost, emergency cooling water is supplied by other pumps, which can be powered by onsite diesel generators. Other safety systems, such as the containment cooling system, also need electric power. PWRs contain between 150–200 fuel assemblies.

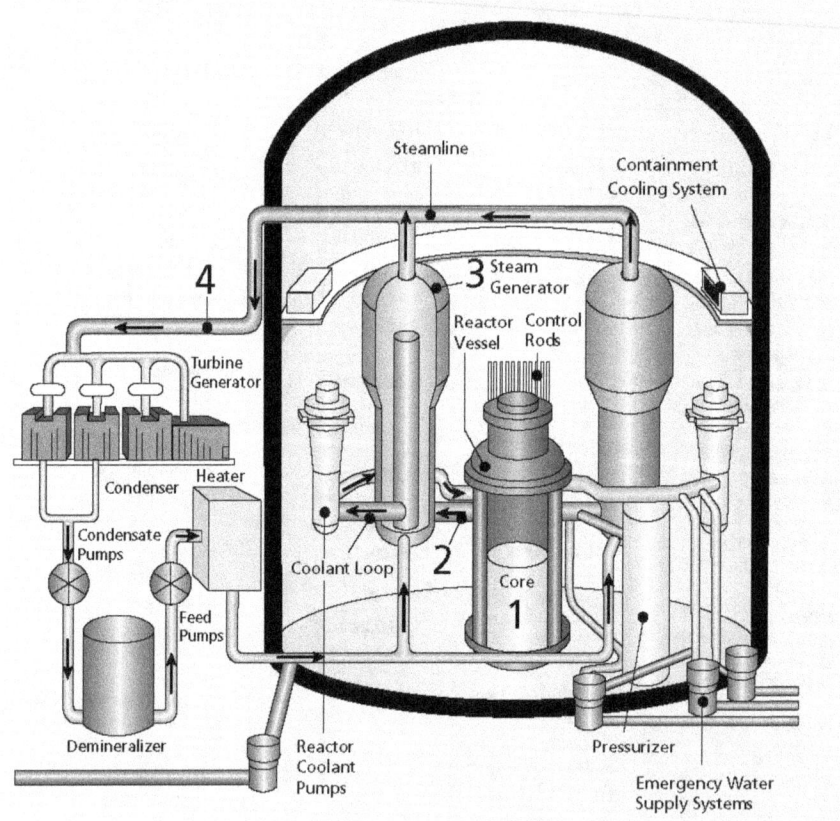

Figure 17. Typical Boiling-Water Reactor

How Nuclear Reactors Work

In a typical design concept of a commercial BWR, the following process occurs:

1. *The core inside the reactor vessel creates heat.*

2. *A steam-water mixture is produced when very pure water (reactor coolant) moves upward through the core, absorbing heat.*

3. *The steam-water mixture leaves the top of the core and enters the two stages of moisture separation where water droplets are removed before the steam is allowed to enter the steamline.*

4. *The steamline directs the steam to the main turbine, causing it to turn the turbine generator, which produces electricity.*

The unused steam is exhausted to the condenser, where it is condensed into water. The resulting water is pumped out of the condenser with a series of pumps, reheated, and pumped back to the reactor vessel. The reactor's core contains fuel assemblies that are cooled by water circulated using electrically powered pumps. These pumps and other operating systems in the plant receive their power from the electrical grid. If offsite power is lost, emergency cooling water is supplied by other pumps, which can be powered by onsite diesel generators. Other safety systems, such as the containment cooling system, also need electric power. BWRs contain between 370–800 fuel assemblies.

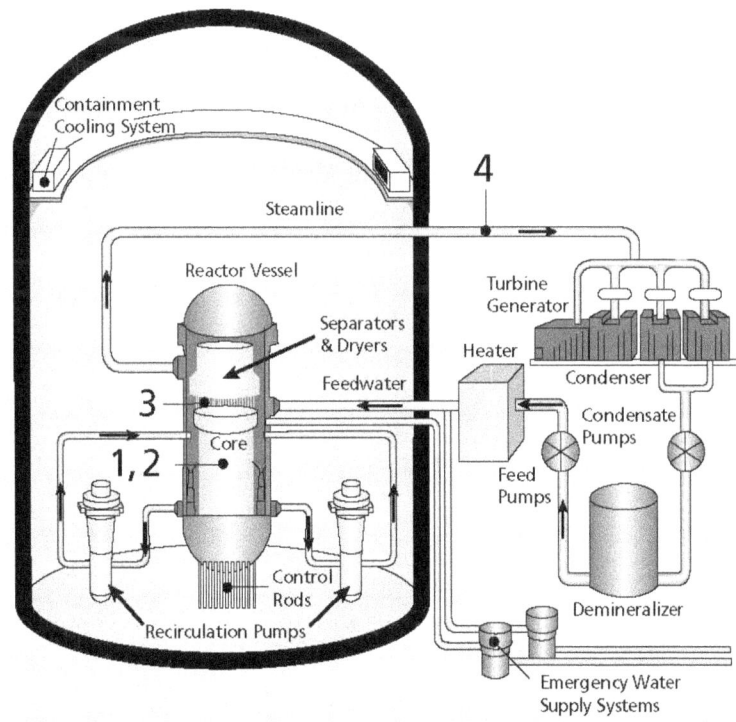

Figure 18. U.S. Commercial Nuclear Power Reactor Operating Licenses—Issued by Year

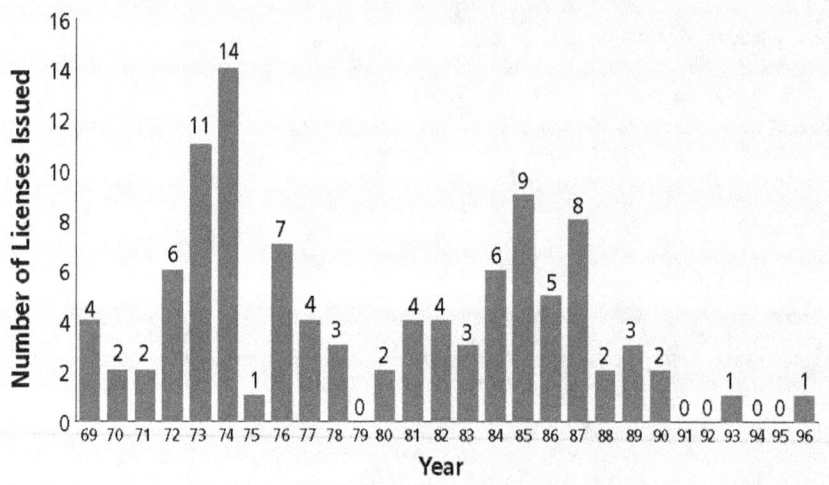

Note: No licenses were issued after 1996.

Table 5. U.S. Commercial Nuclear Power Reactor Operating Licenses—Issued by Year

1969	Dresden 2	**1974**	Arkansas Nuclear 1	**1978**	Arkansas Nuclear 2		Palo Verde 1
	Ginna		Browns Ferry 2		Hatch 2		River Bend 1
	Nine Mile Point 1		Brunswick 2		North Anna 1		Waterford 3
	Oyster Creek		Calvert Cliffs 1	**1980**	North Anna 2		Wolf Creek 1
1970	Point Beach 1		Cooper		Sequoyah 1	**1986**	Catawba 2
	Robinson 2		Cook 1	**1981**	Farley 2		Hope Creek 1
1971	Dresden 3		Duane Arnold		McGuire 1		Millstone 3
	Monticello		FitzPatrick		Salem 2		Palo Verde 2
1972	Palisades		Hatch 1		Sequoyah 2		Perry 1
	Pilgrim		Oconee 3	**1982**	LaSalle 1	**1987**	Beaver Valley 2
	Quad Cities 1		Peach Bottom 3		San Onofre 2		Braidwood 1
	Quad Cities 2		Prairie Island 1		Summer		Byron 2
	Surry 1		Prairie Island 2		Susquehanna 1		Clinton
	Turkey Point 3		Three Mile Island 1	**1983**	McGuire 2		Harris 1
1973	Browns Ferry 1	**1975**	Millstone 2		San Onofre 3		Nine Mile Point 2
	Fort Calhoun	**1976**	Beaver Valley 1		St. Lucie 2		Palo Verde 3
	Indian Point 2		Browns Ferry 3	**1984**	Callaway		Vogtle 1
	Kewaunee		Brunswick 1		Columbia	**1988**	Braidwood 2
	Oconee 1		Calvert Cliffs 2		Diablo Canyon 1		South Texas Project 1
	Oconee 2		Indian Point 3		Grand Gulf 1	**1989**	Limerick 2
	Peach Bottom 2		Salem 1		LaSalle 2		South Texas Project 2
	Point Beach 2		St. Lucie 1		Susquehanna 2		Vogtle 2
	Surry 2	**1977**	Crystal River 3	**1985**	Byron 1	**1990**	Comanche Peak 1
	Turkey Point 4		Davis-Besse		Catawba 1		Seabrook 1
	Vermont Yankee		D.C. Cook 2		Diablo Canyon 2	**1993**	Comanche Peak 2
			Farley 1		Fermi 2	**1996**	Watts Bar 1
					Limerick 1		

Note: Limited to reactors licensed to operate. Year is based on the date the initial full-power operating license was issued. NRC-abbreviated reactor names listed.

Experience

By the end of 2010, U.S. currently operating reactors accumulated nearly 3,100 years of operational experience. Reactors that are permanently shut down account for an additional 385 years of experience (see Figure 18 and Table 5).

Principal Licensing, Inspection, and Enforcement Activities

The NRC conducts a variety of licensing and inspection activities:

- The NRC is reviewing an operating license application from the Tennessee Valley Authority for the Watts Bar Unit 2 reactor under construction near Spring City, TN.

- Typically, each power reactor requests about 10 separate license changes each year. The NRC completed more than 1,000 separate reviews in FY 2010.

- Currently, there are approximately 4,600 NRC-licensed reactor operators. Each operator must requalify every 2 years and apply for license renewal every 6 years.

- On average, the NRC expended approximately 6,490 hours of inspection-related effort at each operating reactor site during 2010 (see Figure 19).

- The NRC reviews applications for proposed new reactors and is developing an inspection program to oversee construction.

- The NRC reviews approximately 3,000 operating experience items, such as fire protection and access authorization programs, from licensed facilities annually.

Figure 19. NRC Inspection Effort at Operating Reactors, 2010

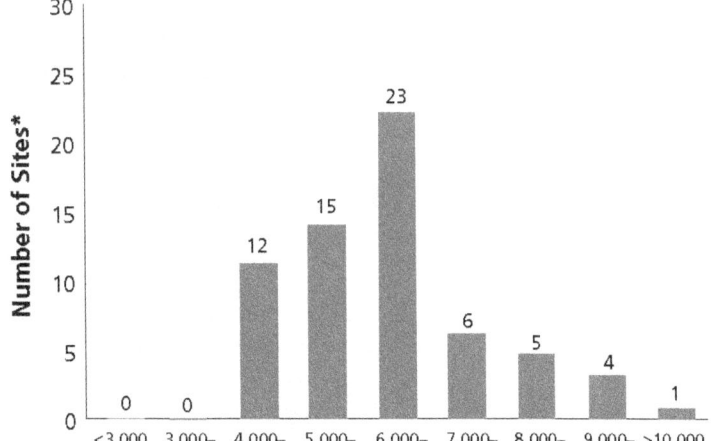

Note: Data include calendar year 2010 hours for all activities related to baseline, plant-specific, generic safety issue, and allegation inspections.

* 66 total sites (Indian Point Units 2 and 3 are treated as separate sites for inspection effort).

- The NRC issues about 15 to 20 escalated enforcement actions per year to operating reactors for violations having a relatively high level of significance with regard to licensed activities affecting public health and safety. The primary enforcement actions, depending on the severity, are notices of violation, civil penalties, and orders.

- The NRC reviews approximately 600 allegations per year; allegations are assertions of impropriety associated with NRC-regulated activities.

- ACRS, an independent body of nuclear, engineering, and safety experts appointed by the Commission, reviews numerous safety issues for existing or proposed reactors and provides independent technical advice to the Commission. ACRS held 10 full Committee meetings and approximately 70 subcommittee meetings during 2010.

- The NRC oversees the decommissioning of 14 nuclear power reactors.

  See Appendix B for permanently shutdown and decommissioning reactors and Appendix N for significant enforcement actions.

OVERSIGHT OF U.S. COMMERCIAL NUCLEAR POWER REACTORS

The NRC does not operate nuclear power plants. Rather, it regulates the operation of the Nation's 104 nuclear power plants by establishing regulatory requirements for their design, construction, and operation. To ensure that the plants are operated safely within these requirements, the NRC licenses the plants to operate, licenses the plant operators, establishes technical specifications for the operation of each plant, and inspects plants daily.

Reactor Oversight Process

The NRC provides continuous oversight of plants through its Reactor Oversight Process (ROP) to verify that they are being operated in accordance with NRC rules, regulations, and license requirements. The NRC has full authority to take action to protect public health and safety, up to and including shutting the plant down.

In general terms, the ROP uses both NRC inspection findings and performance indicators from licensees to assess the safety performance of each plant. The ROP recognizes that issues may have very low to high safety significance, but plants are expected to address all of them effectively. The NRC performs very detailed baseline-level inspections at each plant. If plant problems arise, NRC oversight increases. The agency may perform supplemental inspections and take additional actions to ensure that significant performance issues are addressed. The latest plant-specific inspection findings and performance indicator information can be found on the NRC's Web site (see the Web Link Index).

The ROP takes into account improvements in the performance of the nuclear industry over the past 30 years and improved approaches to inspecting and evaluating the safety performance of NRC-licensed plants. The improvements in plant performance can be attributed both to successful regulatory oversight and to efforts within the nuclear industry.

Resident Inspectors

The NRC has at least two full-time inspectors at each nuclear power plant site to ensure that facilities are meeting NRC regulations.

Figure 20. Industry Performance Indicators: Annual Industry Averages FYs 2001–2010—for 104 Plants

Collective Radiation Exposure

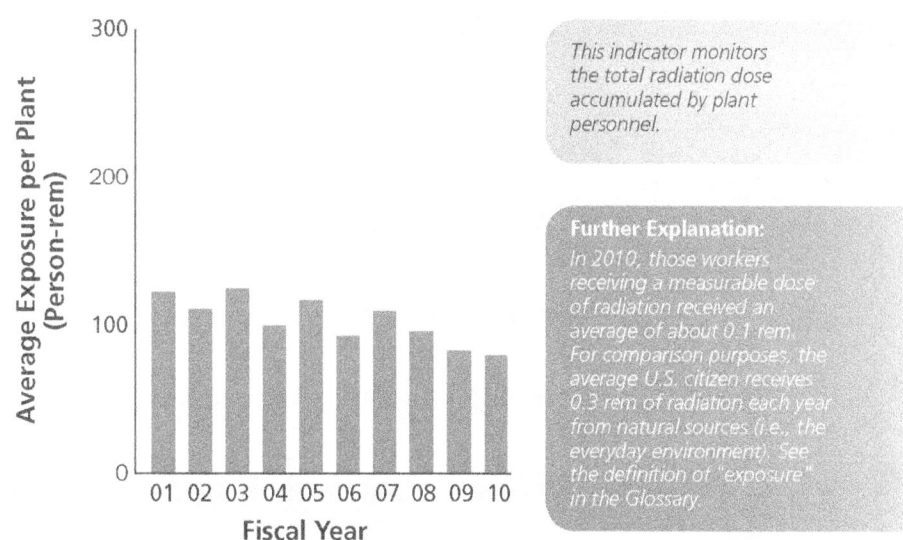

This indicator monitors the total radiation dose accumulated by plant personnel.

Further Explanation:
In 2010, those workers receiving a measurable dose of radiation received an average of about 0.1 rem. For comparison purposes, the average U.S. citizen receives 0.3 rem of radiation each year from natural sources (i.e., the everyday environment). See the definition of "exposure" in the Glossary.

Note: Data represent annual industry averages, with plants in extended shutdown excluded. Data are rounded for display purposes. These data may differ slightly from previously published data as a result of refinements in data quality.

Source: Licensee data as compiled by the NRC

Figure 20. Industry Performance Indicators: Annual Industry Averages FYs 2001–2010—for 104 Plants (continued)

Significant Events

Significant events are events that meet specific NRC criteria, including degradation of safety equipment, a sudden reactor shutdown with complications, or an unexpected response to a sudden degradation of fuel or pressure boundaries. The NRC staff identifies significant events through detailed screening and evaluation of operating experience.

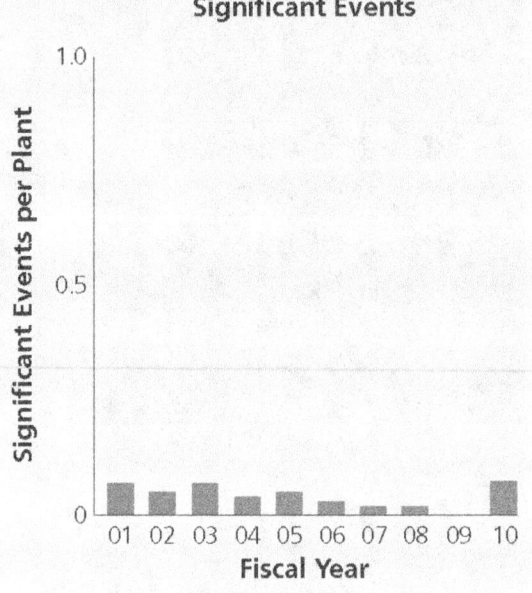

Safety System Failures

Safety system failures are any actual failures, events, or conditions that could prevent a system from performing its required safety function.

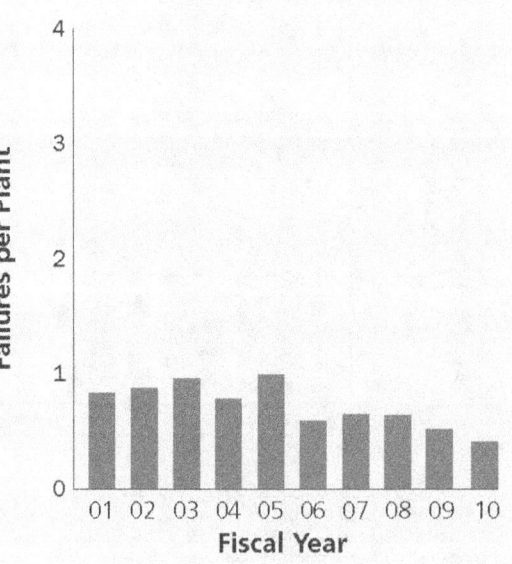

Figure 20. Industry Performance Indicators: Annual Industry Averages FYs 2001–2010—for 104 Plants (continued)

Automatic Scrams While Critical

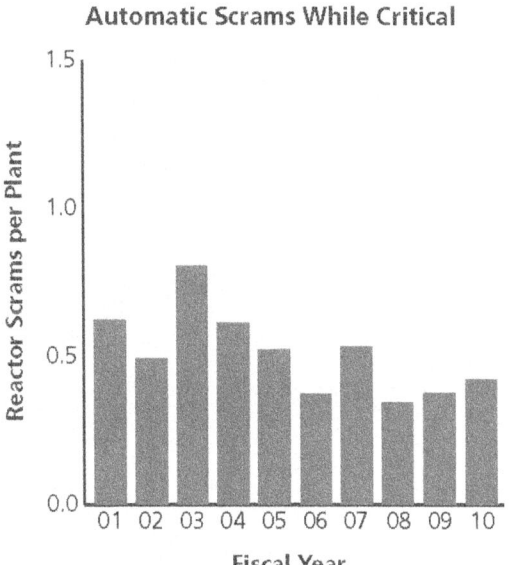

A reactor is said to be "critical" when it achieves a self-sustaining nuclear chain reaction such as when the reactor is operating. The sudden shutting down of a nuclear reactor by the rapid insertion of control rods, either automatically or manually by the reactor operator, is referred to as a "scram." This indicator measures the number of unplanned automatic scrams that occurred while the reactor was critical.

NUCLEAR REACTORS

Safety System Actuations

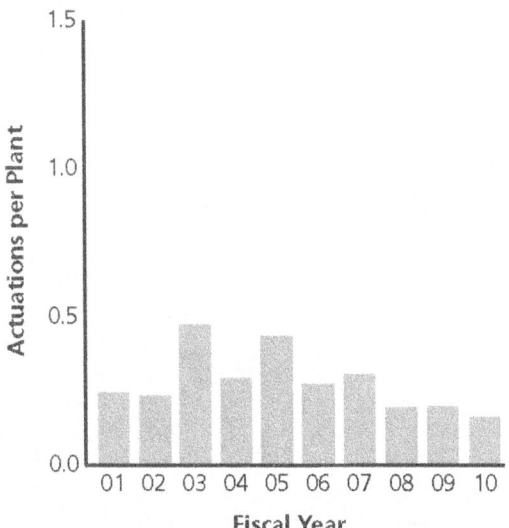

Safety system actuations are certain manual or automatic actions taken to start emergency core cooling systems or emergency power systems. These systems are specifically designed to either remove heat from the reactor fuel rods if the normal core cooling system fails or provide emergency electrical power if the normal electrical systems fail.

Figure 20. Industry Performance Indicators: Annual Industry Averages FYs 2001–2010—for 104 Plants (continued)

The forced outage rate is the number of hours that the plant is unable to operate (forced outage hours) divided by the sum of the hours that the plant is generating and transmitting electricity (unit service hours) and the hours that the plant is unable to operate (forced outage hours).

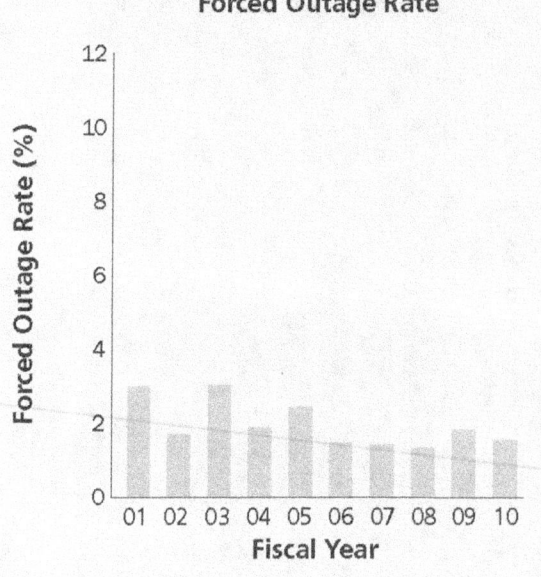

Forced Outage Rate

This indicator is the number of times the plant is forced to shut down because of equipment failures for every 1,000 hours that the plant is in operation and transmitting electricity.

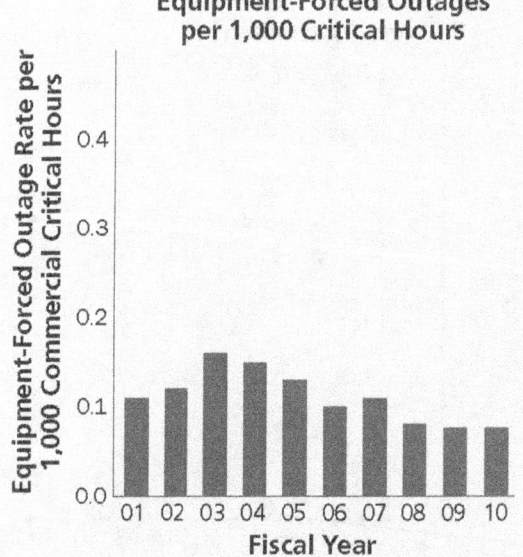

Equipment-Forced Outages per 1,000 Critical Hours

The ROP is described on the NRC's Web site and in NUREG-1649, Revision 4, "Reactor Oversight Process," issued December 2006.

Industry Performance Indicators

In addition to evaluating the performance of each individual plant, the NRC compiles data on overall reactor industry performance using various industry-level performance indicators (see Figure 20).

See Appendix G for the industry performance indicators, which provide additional data for assessing trends in overall industry performance.

NEW COMMERCIAL NUCLEAR POWER REACTOR LICENSING

The NRC is reviewing new reactor applications using a licensing process that substantially improved the system used through the 1990s (see Figure 21). The NRC expects to review 20 combined construction and operating license (called a combined license or COL) applications for approximately 28 new reactors over the next several years and has in place the infrastructure and staff to support the necessary technical work (see Table 6, Figure 22, and the Web Link Index).

Construction and Operating License Applications

As of June 30, 2011, the NRC has received 18 COL applications for 28 new reactor units:

- Calvert Cliffs (MD)
- South Texas Project (TX)
- Bellefonte (AL)
- North Anna (VA)
- William States Lee III (SC)
- Shearon Harris (NC)
- Grand Gulf (MS)
- Vogtle (GA)
- V.C. Summer (SC)

Figure 21. New Reactor Licensing Process

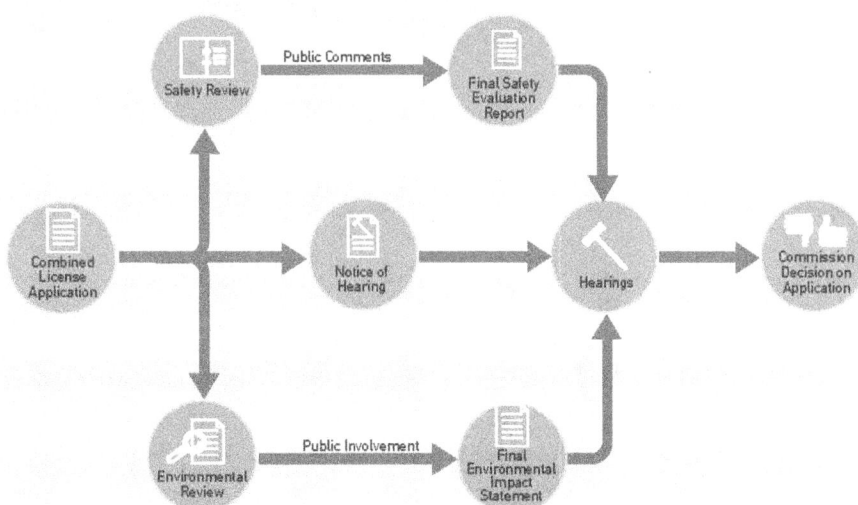

Table 6. U.S. New Nuclear Power Plant Applications

Company (Project/Docket #)	Date of Application	Design	Date Accepted	Site Under Consideration	State	Existing Op. Plant
Calendar Year (CY) 2007 Applications						
NRG Energy (52-012/013)	9/20/07	ABWR	11/29/07	South Texas Project (2 units)	TX	Y
NuStart Energy (52-014/015)	10/30/07	AP1000	1/18/08	Bellefonte (2 units)	AL	N
UNISTAR (52-016)	7/13/07 (Env.), 3/13/08 (Safety)	EPR	1/25/08 6/3/08	Calvert Cliffs (1 unit)	MD	Y
Dominion (52-017)*	11/27/07	US-APWR	1/28/08	North Anna (1 unit)	VA	Y
Duke (52-018/019)	12/13/07	AP1000	2/25/08	William Lee Nuclear Station (2 units)	SC	N
2007 TOTAL NUMBER OF APPLICATIONS = 5 TOTAL NUMBER OF UNITS = 8						
CY 2008 Applications						
Progress Energy (52-022/023)	2/19/08	AP1000	4/17/08	Harris (2 units)	NC	Y
NuStart Energy (52-024)	2/27/08	ESBWR	4/17/08	Grand Gulf (1 unit)	MS	Y
Southern Nuclear Operating Co. (52-025/026)	3/31/08	AP1000	5/30/08	Vogtle (2 units)	GA	Y
South Carolina Electric & Gas (52-027/028)	3/31/08	AP1000	7/31/08	Summer (2 units)	SC	Y
Progress Energy (52-029/030)	7/30/08	AP1000	10/6/08	Levy County (2 units)	FL	N
Detroit Edison (52-033)	9/18/08	ESBWR	11/25/08	Fermi (1 unit)	MI	Y
Luminant Power (52-034/035)	9/19/08	US-APWR	12/2/08	Comanche Peak (2 units)	TX	Y
Entergy (52-036)	9/25/08	ESBWR	12/4/08	River Bend (1 unit)	LA	Y
AmerenUE (52-037)	7/24/08	EPR	12/12/08	Callaway (1 unit)	MO	Y
UNISTAR (52-038)	9/30/08	EPR	12/12/08	Nine Mile Point (1 unit)	NY	Y
PPL Generation (52-039)	10/10/08	EPR	12/19/08	Bell Bend (1 unit)	PA	Y
2008 TOTAL NUMBER OF APPLICATIONS = 11 TOTAL NUMBER OF UNITS = 16						
CY 2009 Applications						
Florida Power & Light Co. (52-040/041)	6/30/09	AP1000	9/4/09	Turkey Point (2 units)	FL	Y
2009 TOTAL NUMBER OF APPLICATIONS = 1 TOTAL NUMBER OF UNITS = 2						
CY 2010 Applications						
No COL applications expected in CY 2010.						
2010 TOTAL NUMBER OF APPLICATIONS = 0 TOTAL NUMBER OF UNITS = 0						
CY 2011 Applications						
No COL applications expected in CY 2011.						
2011 TOTAL NUMBER OF APPLICATIONS = 0 TOTAL NUMBER OF UNITS = 0						
CY 2012 Applications						
No COL applications expected in CY 2012.						
2012 TOTAL NUMBER OF APPLICATIONS = 0 TOTAL NUMBER OF UNITS = 0						
CY 2013 Applications						
Two COL applications are expected in fourth quarter of CY 2013.						
2013 TOTAL NUMBER OF APPLICATIONS = 2 TOTAL NUMBER OF UNITS = 2						
2007–2013 TOTAL NUMBER OF APPLICATIONS = 20 TOTAL NUMBER OF UNITS = 28						

▨ – Accepted/Docketed   ▢ – Expected

Note: Application updates in this table do not show all projects previously mentioned because of changes in intent status or conversion to an early site permit from a COL application. Data as of June 30, 2011.

* Design technology changed by the applicant on June 28, 2010.

Figure 22. New Reactor Licensing Schedule of Applications by Design
Estimated Schedules by Calendar Year (as of June 30, 2011)

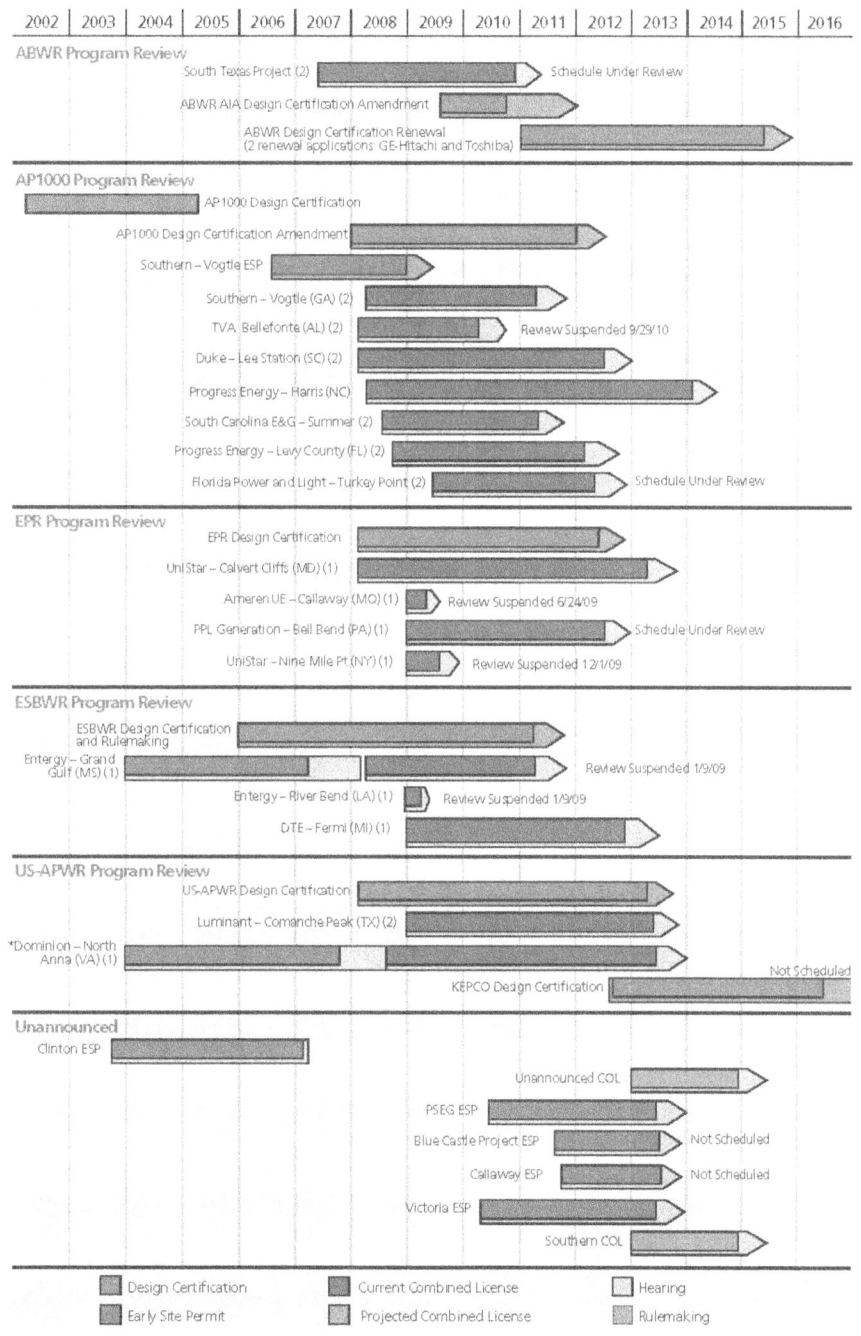

NUCLEAR REACTORS

* Design technology changed by the applicant on June 28, 2010

Figure 22. New Reactor Licensing Schedule of Applications by Design (continued)

Estimated Schedules by Calendar Year (as of June 30, 2011)

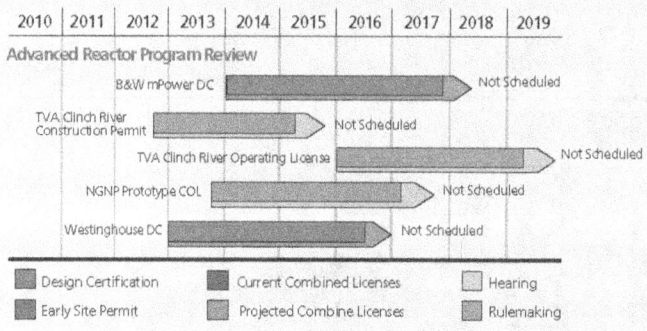

Note: Lines depict approximate dates on schedule. Data on projected applications are based on information from potential applicants and are subject to change. Schedules depicted for future activities represent nominal assumed review durations based on submittal time frames in letters of intent from prospective applicants. Numbers in ( ) next to the COL name indicate the number of units per site. The acceptance review is included at the beginning of the COL review. The rules in 10 CFR Part 2, "Rules of Practice for Domestic Licensing Proceedings and Issuance of Orders," govern hearings on COLs.

- Callaway (MO)
- Levy County (FL)
- Victoria County Station (TX)
- Fermi (MI)
- Comanche Peak (TX)
- River Bend (LA)
- Nine Mile Point (NY)
- Bell Bend (PA)
- Turkey Point (FL)

The NRC suspended or cancelled six COL application reviews due to changes in applicant business strategies (Grand Gulf, Callaway, Nine Mile Point, River Bend, Victoria County Station, and Bellefonte). As of June 2011, the NRC had 12 COL applications for 20 units under active review. Figure 23 shows the locations of the expected new reactor sites.

The staff expects to receive two additional COL applications by the end of CY 2013. For the current review schedule for reactor licensing applications, consult the NRC public Web site (see the Web Link Index).

Public Involvement

The NRC's new reactor licensing process offers many opportunities for public participation. Before it receives an application, the agency uses public meetings to talk to residents in the community near the location where a proposed new reactor may be built to explain how the NRC reviews an application and how the public may participate in the process. Next, the NRC listens to comments on which factors should be considered in the agency's environmental review of the application. The public may then comment on the NRC's draft environmental evaluation that is posted on the agency's Web site. In addition, the public is afforded the opportunity to legally challenge a license application through Atomic Safety and Licensing Board hearings that are announced in press releases and posted on the NRC Web site.

Figure 23. Locations of Applied-for New Nuclear Power Reactors

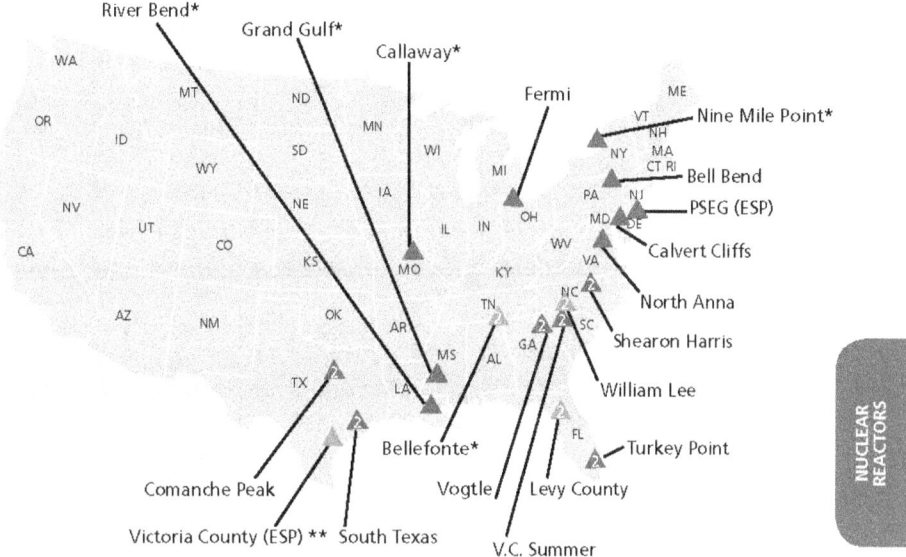

▲ A proposed new reactor at or near an existing nuclear plant

▲ A proposed reactor at a site that has not previously produced nuclear power

▲ = 1 unit   ▲2 = 2 units

*Review suspended.
**COL application amended by applicant to ESP on March 25, 2010.
Note: Data as of June 30, 2011.

The NRC has tailored its new reactor licensing activities to review new applications effectively and efficiently.

Early Site Permits

An early site permit (ESP) provides for early resolution of site safety, environmental protection, and emergency preparedness issues independent of a specific nuclear plant review. The ACRS reviews those portions of the ESP application that concern safety. Mandatory adjudicatory hearings associated with the ESPs are conducted after the completion of the NRC staff's technical review.

The NRC has issued ESPs to the following applicants:

- System Energy Resources, Inc. (Entergy), for the Grand Gulf site in Mississippi

- Exelon Generation Company, LLC, for the Clinton site in Illinois

- Dominion Nuclear North Anna, LLC, for the North Anna site in Virginia

- Southern Nuclear Operating Company, for the Vogtle site in Georgia (includes a limited work authorization)

On March 25, 2010, Exelon Nuclear Texas Holdings (Exelon) submitted an ESP application for the Victoria County Station site located in Victoria County, TX. The ESP application does not include a request for limited work authorization at this time. Exelon previously submitted a COL application for the Victoria County Station site on September 2, 2008, and requested that the COL application be withdrawn when the NRC formally accepts the Victoria County Station ESP application. On June 7, 2010, the NRC docketed the Victoria County ESP application.

NRC staff conducts a vendor inspection at the Tioga Pipe Supply Co., Inc., plant.

NRC staff participates in a site inspection of the proposed new plant in Levy County, FL.

PSEG Power, LLC, and PSEG Nuclear, LLC (PSEG), submitted an ESP application in May 2010 on a site located near the Hope Creek/Salem site. The NRC expects to receive two additional ESP applications by 2012.

Design Certifications

The NRC has issued design certifications (DCs) for four reactor designs that can be referenced in an application for a nuclear power plant. A DC is valid for 15 years from the date of issuance, but it can be renewed for an additional 15 years. The new reactor designs incorporate new elements such as passive safety systems and simplified system designs. These four designs are as follows:

- General Electric-Hitachi Nuclear Energy's (GEH's) Advanced Boiling-Water Reactor (ABWR)
- Westinghouse's System 80+
- Westinghouse's AP600
- Westinghouse's AP1000

The NRC is currently reviewing the following DC applications:

- AREVA's U.S. Evolutionary Power Reactor (EPR)
- Mitsubishi Heavy Industries' U.S. Advanced Pressurized-Water Reactor (US-APWR)

As of June 30, 2011, the NRC completed the technical reviews on a DC application and two DC amendments:

- GEH's Economic Simplified Boiling-Water Reactor (ESBWR)
- Westinghouse's AP1000 DC amendment

- STP Nuclear Operating Company's ABWR DC amendment to address the aircraft impact rule

The NRC expects to complete rulemaking activities for these applications by the end of 2012.

Design Certification Renewals

The NRC received two DC renewal applications for the ABWR from GEH and Toshiba in 2010. Renewals are good for 15 years.

Advanced Reactor Designs

In addition, a range of advanced reactor designs and technologies have emerged that may be submitted to the NRC within the next several years. These technologies include small-sized light-water reactors, liquid-metal reactors, and high-temperature gas-cooled reactors. The NRC will focus its advanced reactor efforts on ensuring that the agency is prepared to address the multiple new technologies being proposed. The NRC has been actively working to develop the regulatory framework in preparation for future licensing application submittals.

New Reactor Construction Inspections

The NRC established a special construction inspection organization in Region II in Atlanta, GA, to inspect licensee construction to ensure that it is performed in compliance with NRC-issued licenses and applicable regulations and to ensure that the as-built facility conforms to its COL. The NRC staff will examine the licensee's operational programs, such as security, radiation protection, and

operator training and qualification, to ensure that the licensee is ready to operate the plant once it is built. The agency's construction site inspectors will verify a licensee's completion of inspections, tests, analyses, and acceptance criteria. The NRC will use these direct inspections and other methods to confirm that the licensee has completed these actions and has met the acceptance criteria included in a COL before allowing startup of the plant.

Starting with the new resident inspectors at the Vogtle site in

Preconstruction activity on limited work authorized at the Vogtle new reactor site.

Preconstruction excavation at the V.C. Summer new reactor site.

April 2010, the NRC will continue to place several full-time inspectors at a site for the duration of the construction phase to oversee day-to-day activities of the licensee and its contractors.

On March 8, 2010, Southern Nuclear Operating Company began site construction at Vogtle Unit 3 under the limited work authorization issued in August 2009. Site activities authorized under the limited work authorization include preliminary construction activities.

The agency also inspects vendor facilities to ensure that products and services furnished to new U.S. reactors meet quality and other regulatory requirements. The NRC has a vendor and quality assurance program and performs quality assurance inspections to ensure that licensees and their contractors meet the regulatory guidelines. To verify compliance with applicable regulations, the NRC inspects domestic and foreign vendors as well as the activities of applicants and licensees.

More information on the NRC's new reactor licensing activities is available on the NRC Web site (see the Web Link Index).

Public Participation in Regulatory Activities

The NRC conducts over 900 public meetings annually and provides opportunities for public involvement in the regulatory process by holding open meetings, conferences, and workshops and issuing rules, regulations, petitions, and technical reports for public comment.

Figure 24. License Renewals Granted for Operating Nuclear Power Reactors

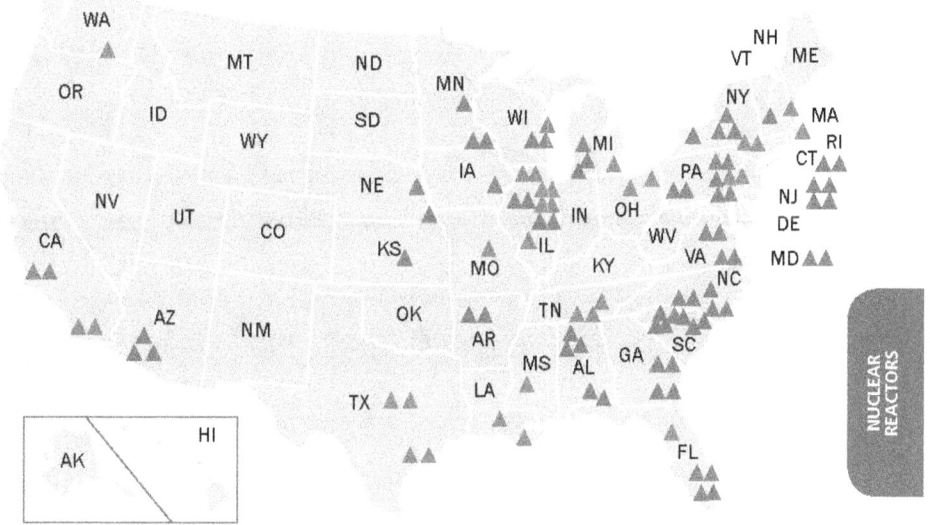

Licensed to Operate (104)
▲ License Renewal Granted (71)
▲ Original License (41)

Note: Data as of July 29, 2011.

REACTOR LICENSE RENEWAL

Based on the Atomic Energy Act of 1954, as amended, the NRC issues licenses for commercial power reactors to operate for 40 years. Under current regulations, licensees may renew their licenses for up to 20 years.

Economic and antitrust considerations, not limitations of nuclear technology, determined the original 40-year term for reactor licenses. However, because of this selected time period, some systems, structures, and components may have been engineered on the basis of an expected 40-year service life.

As of March 2011, approximately two-thirds of the 104 licensed reactor units either have received or are under review for license renewal. Of these, 71 units (at 41 sites) have received renewed licenses (see Figure 24). Figure 25 illustrates the years of commercial operation of operating power reactors. Figure 26 and Table 7 show the expiration dates of operating commercial nuclear licenses.

The decision to seek license renewal rests entirely with nuclear power plant owners and typically is based on the plant's economic situation and on whether it can meet NRC requirements.

Figure 25. U.S. Commercial Nuclear Power Reactors— Years of Operation by the End of 2011

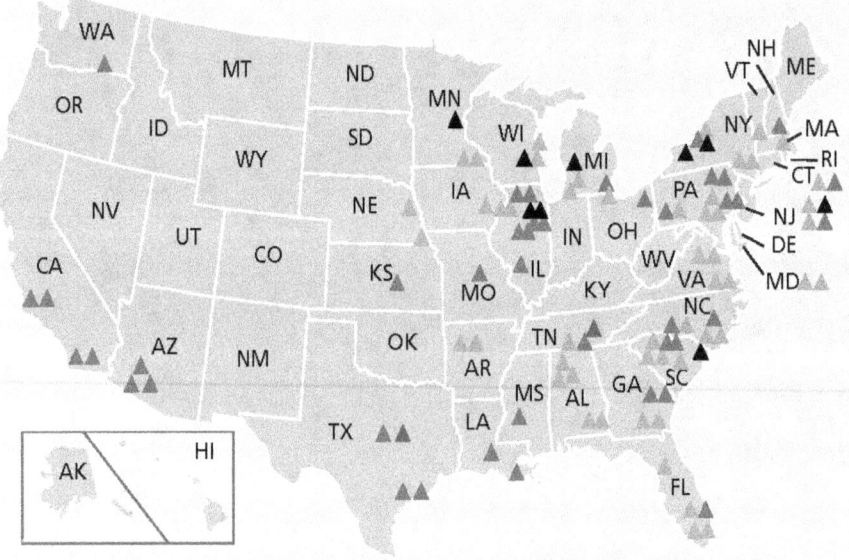

Years of Commercial Operation	Number of Reactors
△ 0–9	0
▲ 10–19	2
▲ 20–29	42
▲ 30–39	51
▲ 40 or more	9

Note: Ages have been rounded up to the end of the year.

The license renewal review process provides continued assurance that the current licensing basis will maintain an acceptable level of safety for the period of extended operation.

The NRC will renew a license only if it determines that a currently operating plant will continue to maintain the required level of safety.

Over the plant's life, this level of safety is enhanced through maintenance of the plant and its licensing basis, with appropriate adjustments to address new information from industry operating experience.

The NRC regulations establish clear requirements for license renewal to ensure safe plant operation for extended plant life, as codified in 10 CFR Part 54, "Requirements for Renewal of Operating Licenses for Nuclear Power Plants." Environmental protection

Figure 26. U.S. Commercial Nuclear Power Reactor Operating Licenses—Expiration by Year

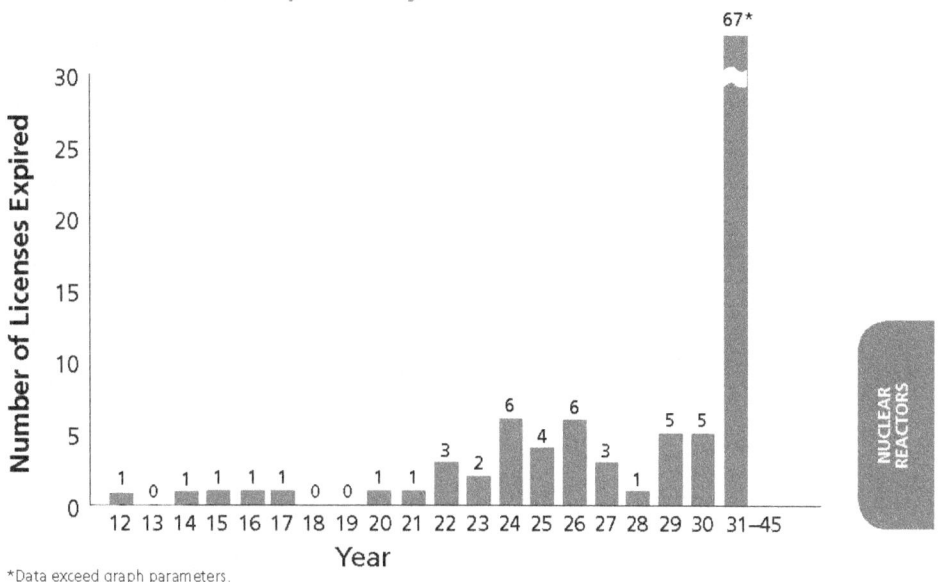

*Data exceed graph parameters.

Table 7. U.S. Commercial Nuclear Power Reactor Operating Licenses—Expiration by Year, 2011–2049

2012	Pilgrim	2027	Braidwood 2		Point Beach 2	2038	Arkansas Nuclear 2
2013	Indian Point 2		South Texas Project 1		Prairie Island 1		Hatch 2
2015	Indian Point 3	2028	South Texas Project 2		Surry 2		North Anna 1
2016	Crystal River 3	2029	Dresden 2		Turkey Point 4	2040	North Anna 2
2017	Davis-Besse		Ginna	2034	Arkansas Nuclear 1		Salem 2
2020	Sequoyah 1		Limerick 2		Browns Ferry 2	2041	Farley 2
2021	Sequoyah 2		Nine Mile Point 1		Brunswick 2		McGuire 1
2022	LaSalle 1		Oyster Creek		Calvert Cliffs 1	2042	Summer
	San Onofre 2	2030	Comanche Peak 1		Cook 1		Susquehanna 1
	San Onofre 3		Monticello		Cooper	2043	Catawba 1
2023	Columbia		Point Beach 1		Duane Arnold		Catawba 2
	LaSalle 2		Robinson 2		Hatch 1		McGuire 2
2024	Byron 1		Seabrook		FitzPatrick		St. Lucie 2
	Callaway	2031	Dresden 3		Oconee 3	2044	Susquehanna 2
	Diablo Canyon 1		Palisades		Peach Bottom 3	2045	Millstone 3
	Grand Gulf 1	2032	Quad Cities 1		Prairie Island 2		Palo Verde 1
	Limerick 1		Quad Cities 2		Three Mile Island 1		Wolf Creek 1
	Waterford 3		Surry 1	2035	Millstone 2	2046	Nine Mile Point 2
2025	Diablo Canyon 2		Turkey Point 3		Watts Bar 1		Harris 1
	Fermi 2		Vermont Yankee	2036	Beaver Valley 1		Hope Creek
	River Bend 1	2033	Browns Ferry 1		Browns Ferry 3		Palo Verde 2
2026	Braidwood 1		Comanche Peak 2		Brunswick 1	2047	Beaver Valley 2
	Byron 2		Fort Calhoun		Calvert Cliffs 2		Palo Verde 3
	Clinton		Kewaunee		St. Lucie 1		Vogtle 1
	Perry		Oconee 1		Salem 1	2049	Vogtle 2
			Oconee 2	2037	Cook 2		
			Peach Bottom 2		Farley 1		

Note: Limited to reactors licensed to operate. NRC-abbreviated reactor names listed. Data as of July 2011.

Figure 27. License Renewal Process

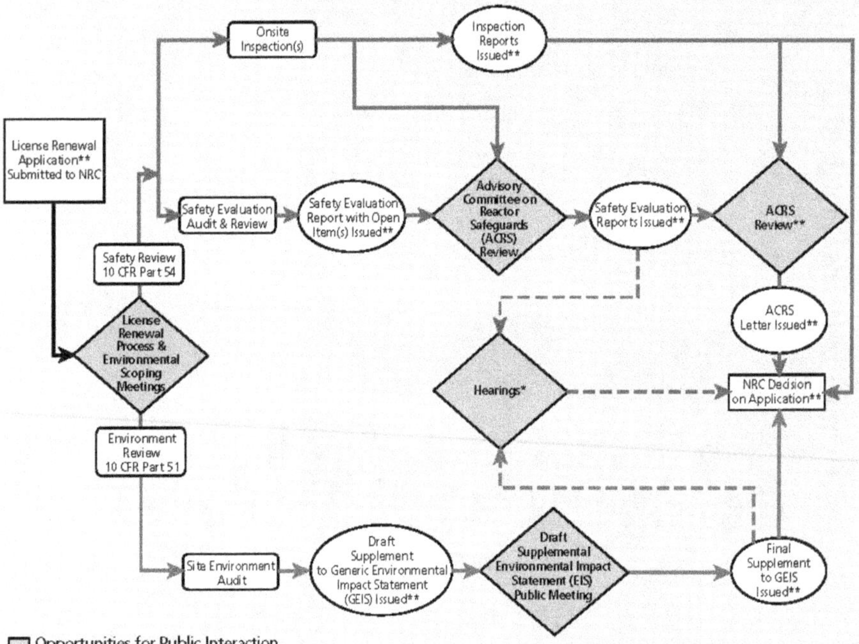

Opportunities for Public Interaction
* If a request for a hearing is granted
** Available at www.nrc.gov

requirements for license renewal are contained in 10 CFR Part 51, "Environmental Protection Regulations for Domestic Licensing and Related Regulatory Functions."

The review of a renewal application proceeds along two paths—one for the review of safety issues and the other for environmental issues (see Figure 27). An applicant must provide the NRC with an evaluation that addresses the technical aspects of plant aging and describes the ways those effects will be managed. The applicant must also prepare for and evaluate the potential impact on the environment if the plant operates for up to an additional 20 years. The NRC reviews the application and verifies the safety evaluation through onsite inspections.

Public Involvement

Public participation is an important part of the license renewal process. Members of the public have several opportunities to question how aging will be managed during the period of extended operation. The NRC makes available to the public information provided by the applicant and holds several public meetings. The agency fully documents its technical and environmental review results and makes them publicly available. In addition, ACRS holds public meetings to discuss technical or safety issues related to plant designs or a particular plant or site. Stakeholder concerns may be litigated in an adjudicatory hearing if any party that would be affected requests a hearing and submits an admissible contention.

For more information, visit the NRC Web site (see the Web Link Index).

RESEARCH AND TEST REACTORS

Nuclear research and test reactors (RTRs) are designed and used for research, testing, and education in nuclear engineering, physics, chemistry, biology, anthropology, medicine, materials sciences, and related fields. These reactors do not produce commercial electricity, but they help prepare people for nuclear-related careers in the fields of nuclear engineering, electric power, national defense, health services, research, and education.

The largest U.S. RTR (at 20 megawatts thermal) is 75 times smaller than the smallest U.S. commercial power nuclear reactor (at 1,500 megawatts thermal). There are 43 licensed RTRs:

- 31 RTRs operating in 22 States (see Figure 28)

- 12 reactors shut down and in various stages of decommissioning

See Appendix E for a list of the 31 operating RTRs regulated by the NRC.

RTRs licensed to operate at a power level of 2 megawatts or greater are

NUCLEAR REACTORS

Figure 28. U.S. Nuclear Research and Test Reactors

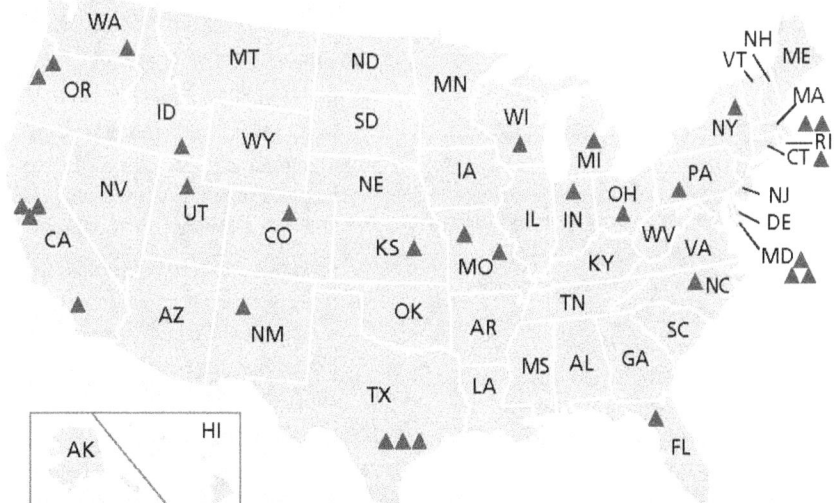

▲ Licensed/Currently Operating (31)

inspected annually. RTRs licensed to operate at power levels below 2 megawatts are inspected every 2 years.

Since 1958, 82 licensed RTRs have been decommissioned.

See Appendix F for a list of the 12 RTRs regulated by the NRC that are in the process of decommissioning.

Principal Licensing and Inspection Activities

The NRC's principal licensing and inspection activities related to RTRs include the following:

- licensing the 31 operating RTRs, including license renewals and license amendments

- licensing approximately 97 RTR operators

- requalifying each operator before renewal of his or her 6-year license

- conducting approximately 36 RTR inspections each year

NUCLEAR REGULATORY RESEARCH

The NRC's research program supports the agency's regulatory mission by providing technical advice, tools, and information to identify and resolve safety issues, make regulatory decisions, and promulgate regulations and guidance. This includes conducting confirmatory experiments and analyses; developing technical bases that support the NRC's safety decisions; and preparing the agency for the future by evaluating the safety aspects of new technologies and designs for nuclear

reactors, materials, waste, and security. The research program focuses on challenges as the industry continues to evolve, including potential new safety issues, management of aging and material degradation issues, technical issues associated with the deployment of new technologies and reactor designs, and retention of technical skills as experienced staff retires.

In the near term, research supports oversight of operating light-water reactors, the technology currently used in the United States. However, recent applications for advanced light-water reactors and preapplication activity regarding nonlight-water reactor vendors have prompted the agency to consider longer term research needs.

The NRC ensures protection of public health, safety, and the environment through research programs that do the following:

- Examine technical areas.

  » material degradation (e.g., stress-corrosion cracking, aging management, degradation mitigation technologies, boric acid corrosion, and embrittlement)

  » new and evolving technologies (e.g., new reactor technology, mixed oxide fuel performance, digital instrumentation and control, and safety-critical software)

  » experience gained from operating reactors

  » probabilistic risk assessment methods

  » seismic and geotechnical hazards

Demonstration of Loss-of-Coolant Accident Experiment

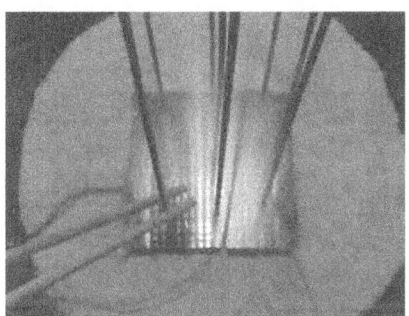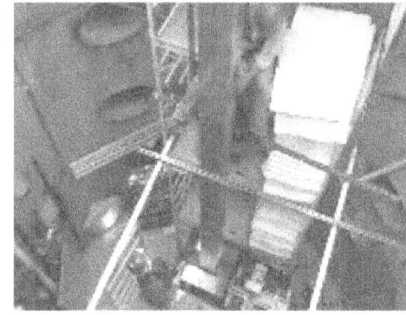

The Sandia Fuel Project gathers data on the behavior of fuel assemblies in storage pools and reactor cores in a complete loss-of-coolant scenario. Electrically heated rods are inserted in commercially available PWR fuel assemblies and a prototypic spent fuel rack. As shown, the assembly is ignited in a zirconium cladding fire and later inspected for damage. The results represent actual fuel assembly responses and are used to support computer modeling codes for accident scenarios.

» ability of equipment to function in a harsh environment (e.g., heat, radiation, humidity)

» structural integrity assessments of reactor component degradation (e.g., nondestructive evaluation techniques and protocols)

• Examine human factors issues, including safety culture and computerization and automation of control rooms.

• Develop and improve computer codes as computational abilities expand and additional experimental and operational data allow for more realistic simulation. These computer codes analyze a wide spectrum of technical areas, including severe accidents, radionuclide transport through the environment, health effects of radioactive releases, nuclear criticality, fire conditions in nuclear facilities, thermal-hydraulic performance of reactors, reactor fuel performance, and nuclear power plant risk assessment.

- Ensure the secure use and management of nuclear facilities and radioactive materials by investigating potential security vulnerabilities and possible compensatory actions.

The NRC dedicates about 7 percent of its personnel and about 15 percent of its contracting funds to research. This research enables the NRC's highly skilled, experienced experts to formulate sound technical solutions based on science and to support timely and realistic regulatory decisions.

The NRC research budget for FY 2011 is approximately $61 million. This includes contracts with national laboratories, universities, and other research organizations for greater expertise and access to research facilities. Figure 29 illustrates the primary areas of research.

The NRC directs about three-fourths of the research program toward maintaining the safety of existing operating reactors. The agency is also directing research in support of regulating new and advanced reactors.

Radioactive waste programs and security are additional focus areas for research. Infrastructure support includes information technology and human resources.

The NRC also has cooperative agreements with universities and nonprofit organizations to research specific areas of interest to the agency.

See Appendix M for a list of cooperative agreements.

The NRC recently asked the National Academies to assess the feasibility of doing a study on the cancer risk for populations around nuclear power facilities. The NRC expects the feasibility study to be complete in the spring of 2012.

Figure 29. NRC Research Funding, FY 2011

Total: $61 Million

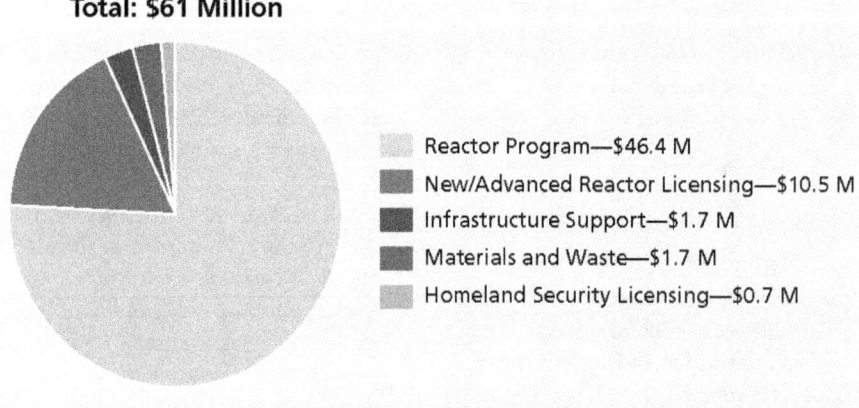

Reactor Program—$46.4 M
New/Advanced Reactor Licensing—$10.5 M
Infrastructure Support—$1.7 M
Materials and Waste—$1.7 M
Homeland Security Licensing—$0.7 M

Note: Totals may not equal sum of components because of independent rounding.

Nuclear Research at Universities

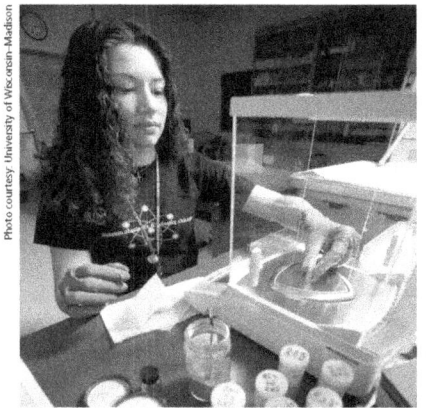 

Universities and other academic institutions use nuclear material in laboratory experiments and to provide health physics support to other institutional nuclear materials users.

The State-of-the-Art Consequence Analysis (SOARCA) research project currently underway will develop realistic estimates of potential health effects from nuclear power plant accident scenarios that could involve releases of radioactive material into the environment. SOARCA improves methods and models for realistically evaluating plant responses during a severe accident.

The NRC collaborates with the international research community on both light-water and nonlight-water reactor technologies. These collaborations enable the agency to better leverage its resources, to initiate activities focused on evolutionary advances in existing technologies, and to determine the safety implications of new technologies. Collaboration is aided by the agency's leadership role in the standing committees and senior advisory groups of international organizations, such as IAEA and NEA.

The NRC also has research agreements with foreign governments for international cooperative research. The NRC is engaged in 100 cooperative research agreements with more than two dozen countries and NEA:

- Halden Reactor Project in Norway. For over 50 years, this collaboration has allowed for research and development of fuel, reactor internals, plant control and monitoring, human factors, and human reliability analysis.

- International Steam Generator Tube Integrity Program with Japan, South Korea, Canada, and others. This longstanding program models and predicts the impact of the aging and materials degradation process on tubing.

NUCLEAR MATERIALS

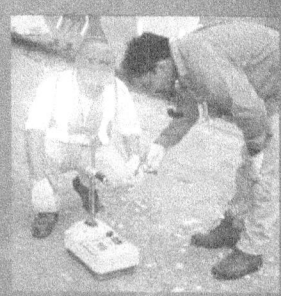

Top: Workers using a moisture density gauge. (Photo courtesy: APNGA)

Middle: A Leskel Gamma Knife® headframe uses radiation beams to treat people with brain cancer. (Photo courtesy: Elekta)

Bottom: Yellowcake is produced by uranium recovery facilities and is then transported to a uranium conversion facility.

The NRC regulates nuclear materials for use in medical, industrial, and academic applications. It also regulates the phases of the nuclear fuel cycle, which begins with the uranium recovery, conversion, enrichment, and fabrication facilities that produce nuclear fuel for power plants.

MATERIALS LICENSES

Through agreements with the NRC, many States have assumed regulatory authority over radioactive materials, with the exception of nuclear reactors, fuel facilities, and certain quantities of special nuclear material. These States are called Agreement States, as shown in gold in Figure 30. The NRC and Agreement State regulatory programs are designed to ensure that licensees

safely use these materials and do not endanger public health and safety or cause damage to the environment.

The NRC and Agreement States have issued approximately 22,000 licenses for general use of nuclear materials (see Table 8):

- The NRC administers approximately 3,000 licenses.

- 37 Agreement States administer approximately 19,000 licenses.

Reactor- and accelerator-produced radionuclides are used extensively throughout the United States for civilian and military industrial applications; basic and applied research; manufacture of consumer products; academic studies; and medical diagnosis, treatment, and research.

Figure 30. Agreement States

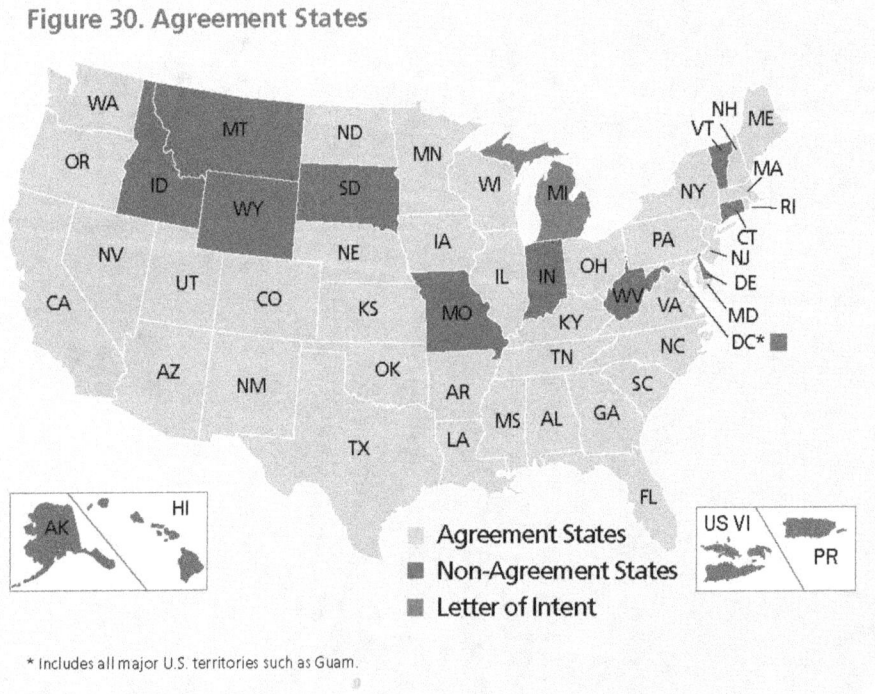

* Includes all major U.S. territories such as Guam.

Table 8. U.S. Materials Licenses by State

	Number of Licenses			Number of Licenses	
State	NRC	Agreement States	State	NRC	Agreement States
Alabama	17	445	Montana	89	0
Alaska	63	0	Nebraska	5	148
Arizona	11	374	Nevada	3	248
Arkansas	6	218	New Hampshire	6	79
California	56	1,913	New Jersey	42	672
Colorado	19	346	New Mexico	15	193
Connecticut	188	0	New York	27	1,441
Delaware	58	0	North Carolina	17	760
District of Columbia	41	0	North Dakota	9	77
Florida	21	1,707	Ohio	42	657
Georgia	17	503	Oklahoma	19	248
Hawaii	41	0	Oregon	4	337
Idaho	79	0	Pennsylvania	59	767
Illinois	32	713	Rhode Island	1	49
Indiana	289	0	South Carolina	15	419
Iowa	4	172	South Dakota	43	0
Kansas	10	301	Tennessee	20	601
Kentucky	8	453	Texas	50	1,647
Louisiana	12	527	Utah	10	198
Maine	2	126	Vermont	35	0
Maryland	80	609	Virginia	64	424
Massachusetts	80	497	Washington	16	421
Michigan	514	0	West Virginia	179	0
Minnesota	12	150	Wisconsin	17	328
Mississippi	6	334	Wyoming	83	0
Missouri	291	0	Others*	153	0
			Total	**2,958**	**19,132**

☐ Agreement State

* Others include major U.S. territories.

Note: The NRC and Agreement State data are the latest available as of March 2011.
The NRC licenses Federal agencies in Agreement States.

NUCLEAR MATERIALS

MEDICAL AND ACADEMIC

In both medical and academic settings, the NRC reviews the facilities, personnel, program controls, and equipment to ensure the safety of the public, patients, and workers who might be exposed to radiation.

Medical

The NRC and Agreement States issue licenses to hospitals and physicians for the use of radioactive materials in medical treatments. In addition, the NRC develops guidance and regulations for use by licensees and maintains a committee of medical experts to obtain advice about the use of radioactive materials in medicine. The NRC regulations require that physicians and physicists have special training and experience to practice radiation medicine. The training emphasizes safe operation of nuclear-related equipment and accurate recordkeeping. The Advisory Committee on the

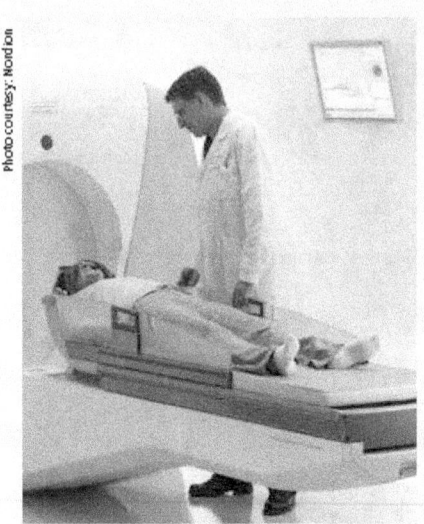

Gamma Knife® used for treating brain tumors.

Medical Uses of Isotopes comprises physicians, scientists, and other health care professionals who advise the NRC staff on initiatives in the medical uses of radioactive materials.

Nuclear Medicine

About one-third of all patients admitted to hospitals are diagnosed or treated using radioactive materials. This branch of medicine is known as nuclear medicine, and the radioactive materials for treatment are called radiopharmaceuticals. Doctors of nuclear medicine use radiopharmaceuticals to diagnose patients through in *vivo* tests (direct administration of radiopharmaceuticals to patients) or in *vitro* tests (the addition of radioactive materials to lab samples taken from patients). Doctors also use radiopharmaceuticals and radiation-producing devices to treat conditions such as hyperthyroidism and certain forms of cancer and to ease

pain caused by bone cancer. In the past decade, the use of nuclear medicine for treatment and diagnoses has increased significantly.

Diagnostic Procedures

For most diagnostic procedures in nuclear medicine, a small amount of radioactive material is administered, either by injection, inhalation, or oral administration. The radiopharmaceutical collects in the organ or area being evaluated, where it emits photons. These photons can be detected by a device known as a gamma camera, which produces images that provide information about the organ function and composition.

Radiation Therapy

The primary objective of radiation therapy is to deliver an accurately prescribed dose of radiation to the target site while minimizing the radiation dose to surrounding healthy tissue. Radiation therapy can be used to treat cancer or to relieve symptoms associated with certain diseases, such as cancer. Treatments often involve multiple exposures spaced over a period of time for maximum therapeutic effect. When used to treat malignant diseases, radiation therapy is often delivered in combination with surgery or chemotherapy.

There are three main categories of radiation therapy:

1. External beam therapy (also called teletherapy) is a beam of radiation directed to the target tissue. There are several different categories of external beam therapy units. The

type of treatment machine that is regulated by the NRC contains a high-activity radioactive source (usually cobalt-60) that emits photons to treat the target site.

2. In brachytherapy treatments, sealed radioactive sources are permanently or temporarily placed near or on a body surface, in a body cavity, directly on a surface within a cavity, or directly on the cancerous tissue. The radiation dose is delivered at a distance of up to an inch (a few centimeters) from the target area.

3. Therapeutic radiopharmaceuticals are quantities of unsealed radioactive materials that localize in a specific region or organ system to deliver a large radiation dose.

Academic

The NRC issues licenses to academic institutions for educational and research purposes. For example, qualified instructors use radioactive materials in classroom demonstrations. Scientists in a wide variety of disciplines use radioactive materials for laboratory research.

INDUSTRIAL

The NRC and Agreement States license users of radioactive material for the specific type, quantity, and location of material that may be used. Radionuclides are used in industrial and commercial applications, including industrial radiography, gauges, well logging, and manufacturing. For example, radiography uses radiation sources to find structural defects in metallic materials and welds. Gas chromatography uses low-energy

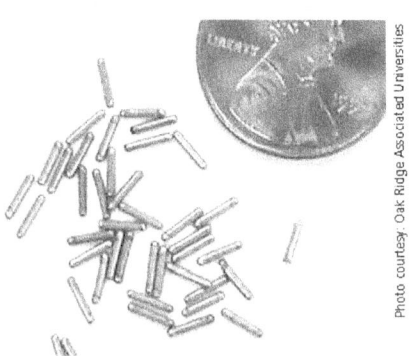

Iodine-125 and palladium-103 found in implantable seeds are primarily used to treat prostate cancer.

radiation sources for identifying the chemical elements in an unknown substance. Gas chromatography can determine the components of complex mixtures, such as petroleum products, smog, and cigarette smoke, and can be used in biological and medical research to identify the components of complex proteins and enzymes. Well-logging devices use a radioactive source and detection equipment to make a record of geological formations down a bore hole. This process is used extensively for oil, gas, coal, and mineral exploration.

Nuclear Gauges

Nuclear gauges are used as nondestructive devices to measure the physical properties of products and industrial processes as a part of quality control. Gauges use radiation sources to determine the thickness of paper products, fluid levels in oil and chemical tanks, and the moisture and density of soils and material at construction sites. There are fixed and portable gauges.

A fixed gauge consists of a radioactive source that is contained in a source

NUCLEAR MATERIALS

holder. When the user opens the container's shutter, a controlled beam of radiation hits the material or product being processed or controlled. A detector mounted opposite the source measures the radiation passing through the product. The gauge readout or computer monitor shows the measurement. The material and process being monitored dictate the selection of the type, energy, and strength of radiation.

Fixed fluid gauges are installed on a pipe that is used by the beverage, food, plastics, and chemical industries to measure the densities, flow rates, levels, thicknesses, and weights of a wide variety of materials and surfaces.

Figure 31 shows a portable gauge configuration in which the gamma source is placed under the surface of the ground through a tube. Radiation is then transmitted directly to the detector on the bottom of the gauge, allowing accurate measurements of compaction. Industry uses such gauges to monitor the structural integrity of roads,

buildings, and bridges and to explore for oil, gas, and minerals. Airport security uses nuclear gauges to detect explosives in luggage at airports.

A portable gauge is a radioactive source and detector mounted together in a portable shielded device. The device is placed on the object to be measured, and the source is either inserted into the object or the gauge relies on a reflection of radiation from the source to bounce back to the bottom of the gauge. The detector in the gauge measures the radiation either directly from the inserted source or from the reflected radiation.

The radiation measurement indicates the thickness, density, moisture content, or some other property that is displayed on a gauge readout or on a computer monitor. The top of the gauge has sufficient shielding to protect the operator while the source is exposed. When the measuring process is completed, the source is retracted or a shutter closes, minimizing exposure from the source.

Figure 31. Moisture Density Gauge

Direct Transmission

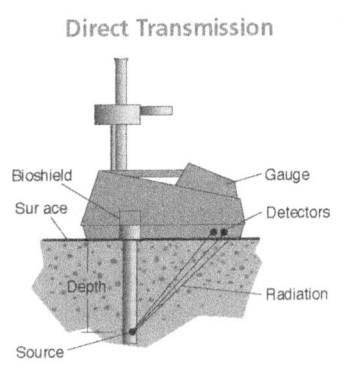

A moisture density gauge indicates whether a foundation is suitable for constructing a building or roadway.

Figure 32. Commercial Irradiator

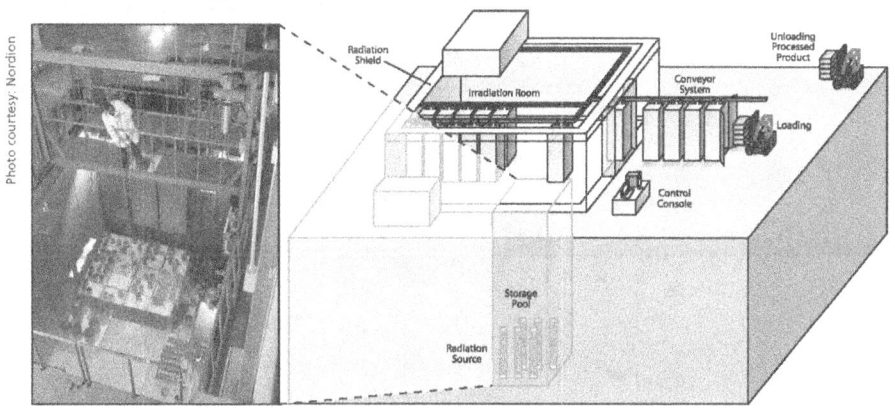

Photo courtesy: Nordion

Commercial Irradiators

Commercial irradiators expose products such as food, food containers, spices, medical supplies, and wood flooring to radiation to eliminate harmful bacteria, germs, and insects or for hardening or other purposes (see Figure 32). The gamma radiation does not leave any radioactive residue or cause any of the treated products to become radioactive themselves. The source of that radiation can be radioactive materials (e.g., cobalt-60), an x-ray tube, or an electron beam.

The NRC and Agreement States license approximately 50 commercial irradiators nationwide. For the past 40 years, the U.S. Food and Drug Administration and other agencies have approved the irradiation of meat and poultry as well as other foods, including fresh fruits, vegetables, and spices. The amount of radioactive material in the devices can range from 1 curie to 10 million curies. NRC regulations protect workers and the public from radiation involved in irradiation operations.

Generally, two types of commercial irradiators are in operation in the United States: underwater and wet-source-storage panoramic models.

In the case of underwater irradiators, the sealed sources (radioactive material encased inside a capsule) that provide the radiation remain in the water at all times, providing shielding for workers and the public. The product to be irradiated is placed in a watertight container, lowered into the pool, irradiated, and then removed.

With wet-source-storage panoramic irradiators, the radioactive sealed sources are also stored in the water, but they are raised into the air to irradiate products that are automatically moved in and out of the room on a conveyor system. Sources are then lowered back to the bottom of the pool. For this type of irradiator, thick concrete walls or steel barriers protect workers and the public when the sources are lifted from the pool.

NUCLEAR MATERIALS

Figure 33. Life Cycle Approach to Source Security

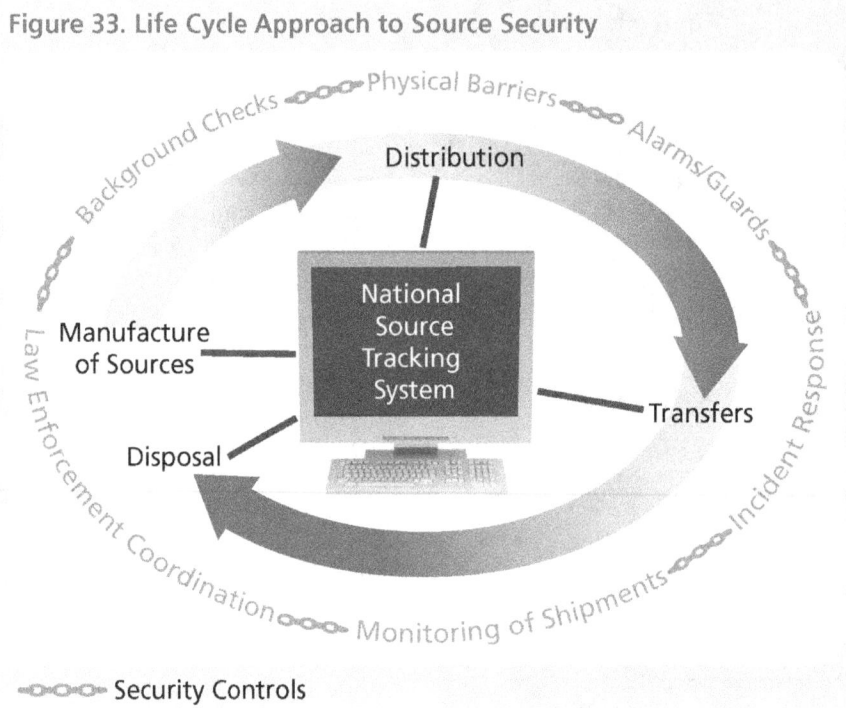

Security Controls

TRANSPORTATION

About 3 million packages of radioactive materials are shipped each year in the United States, either by road, rail, air, or water. This represents less than 1 percent of the Nation's yearly hazardous material shipments. Regulating the safety of commercial radioactive material shipments is the joint responsibility of the NRC and the U.S. Department of Transportation (DOT).

The vast majority of these shipments consist of small amounts of radioactive materials used in industry, research, and medicine. The NRC requires such materials to be shipped in accordance with DOT's hazardous materials transportation safety regulations.

MATERIAL SECURITY

In January 2009, the NRC deployed its National Source Tracking System (NSTS), by which the agency and its Agreement States track the manufacture, distribution, and ownership of the most high-risk sources. Licensees use the NSTS, a secure Web-based system, to enter up-to-date information on the receipt or transfer of tracked radioactive sources (see Figure 33).

Over the past several years, the NRC and the Agreement States have increased the controls they have imposed on the most sensitive radioactive materials, including physical security requirements and limited personnel access to the materials.

Figure 34. The Nuclear Fuel Cycle

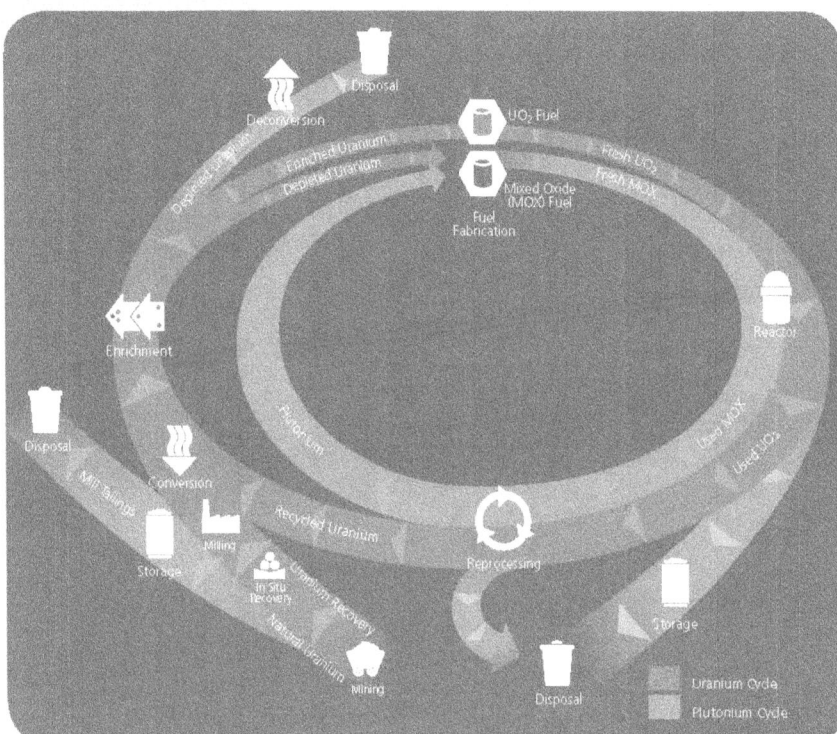

Working with other Federal agencies, such as DHS, the NRC has also implemented a voluntary program of additional security improvements. Together, these activities help make potentially dangerous radioactive sources even more secure and less vulnerable to terrorists.

Principal Licensing and Inspection Activities

Each year, the NRC issues approximately 2,700 new licenses, license renewals, and amendments for existing material licenses.

The NRC conducts approximately 1,250 health and safety and security inspections of its nuclear materials licensees each year.

URANIUM RECOVERY

Figure 34 illustrates the nuclear fuel cycle, which begins with the uranium recovery, conversion, enrichment, and fabrication facilities that produce nuclear fuel for power plants. To make fuel for reactors, uranium is recovered or extracted from the ore, converted, enriched, and manufactured into fuel pellets.

Figure 35. The In Situ Uranium Recovery Process

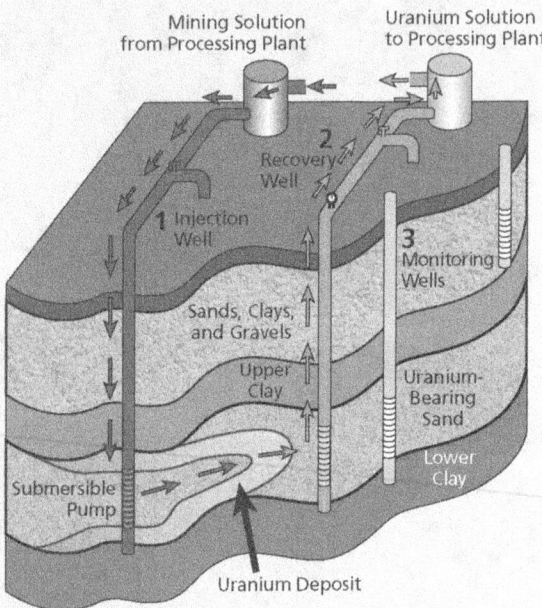

Injection wells (1) pump a chemical solution—typically sodium bicarbonate, hydrogen peroxide, and oxygen—into the layer of earth containing uranium ore. The solution dissolves the uranium from the deposit in the ground and is then pumped back to the surface through recovery wells (2) and sent to the processing plant to be converted into uranium yellowcake. Monitoring wells (3) are checked regularly to ensure that uranium and chemicals are not escaping from the drilling area.

The NRC does not regulate traditional mining, but it does regulate the processing of uranium ore It has jurisdiction over uranium recovery facilities such as conventional mills and in situ recovery (ISR) facilities.

The NRC has a well-established regulatory framework for ensuring that uranium recovery facilities are appropriately licensed, operated, decommissioned, and monitored to protect public health and safety.

Conventional Uranium Mill

A conventional uranium mill is a chemical plant that extracts uranium from mined ore. Conventional mills are typically located in areas of low population density, within about 50 kilometers (30 miles) of a uranium mine. The mined ore is transported to the mill, where it is crushed. Sulfuric acid then dissolves the soluble components, including 90 to 95 percent of the uranium, from the ore. The uranium is then separated from the solution, concentrated, and dried to form yellowcake (yellow uranium oxide powder). There are 16 uranium recovery sites licensed by the NRC. Of these, 10 are in various stages of decommissioning and one is in standby status with the potential to restart in the future.

In Situ Recovery

ISR is another means of extracting uranium—this time from underground ore. ISR facilities recover uranium from ores for which recovery may

Figure 36. Locations of NRC-Licensed Uranium Recovery Facility Sites

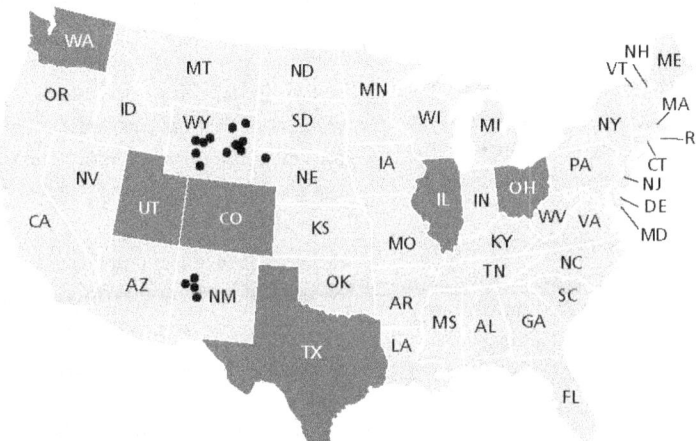

● NRC-licensed uranium recovery facility sites (16)

■ States with authority to license uranium recovery facility sites

States where the NRC has retained authority to license uranium recovery facilities

Table 9. Locations of NRC-Licensed Uranium Recovery Facilities

Licensee	Site Name, Location
In Situ Recovery Facilities	
Uranium One	Willow Creek, WY
Crow Butte Resources, Inc.	Crow Butte, NE*
Hydro Resources, Inc.°	Crownpoint, NM
Power Resources, Inc.	Smith Ranch and Highlands, WY*
Uranium One	Moore Ranch, WY
Conventional Uranium Mill Recovery Facilities	
American Nuclear Corp.†	Gas Hills, WY
Bear Creek Uranium Co.†	Bear Creek, WY
Exxon Mobil Corp.†	Highlands, WY
Homestake Mining Co.†	Homestake, NM
Kennecott Uranium Corp.°	Sweetwater, WY
Pathfinder Mines Corp.†	Lucky Mc, WY
Pathfinder Mines Corp.†	Shirley Basin, WY
Rio Algom Mining, LLC†	Ambrosia Lake, NM
Umetco Minerals Corp.†	Gas Hills, WY
United Nuclear Corp.†	Church Rock, NM
Western Nuclear, Inc.†	Split Rock, WY

Note: For further details on NRC-related uranium recovery facility applications in review and applications, restarts, and expansions, see the Web Link Index. This table does not include uranium recovery facilities licensed by Agreement States.

* Satellite facilities are located within the State.

† These sites are undergoing decommissioning.

° Kennecott has an operating license but is in "standby" mode. Hydro has an operating license, but the facility has not yet been constructed.

not be economically viable by other methods. In this process, a solution of native ground water typically mixed with oxygen or hydrogen peroxide and sodium bicarbonate or carbon dioxide is injected through wells into the ore to dissolve the uranium. The resulting solution is pumped from the rock formation, and the uranium is then separated from the solution to form yellowcake (see Figure 35). The United States has about 12 such ISR facilities. Of these facilities, the NRC licenses five, and Agreement States license the rest (see Figure 36 and Table 9).

Because of the resurgence of interest in the construction of new nuclear power plants, the agency anticipates as many as 27 applications for new uranium recovery facilities and expansions or restarts of existing facilities in the next few years. As of March 2011, the agency received seven applications for new facilities and four applications to expand or restart an existing facility. The current status of applications can be found on the NRC's Web site (see the Web Link Index). Existing facilities and new potential sites are located in Wyoming, New Mexico, Nebraska, South Dakota, and Nevada, and in the Agreement States of Texas, Colorado, and Utah (see Figure 37 and Table 10). The NRC works closely with stakeholders, including Native American Tribal governments, to address concerns with the licensing of new uranium recovery facilities.

The NRC is also responsible for the following:

- inspecting and overseeing both active and inactive uranium recovery facilities

- ensuring that siting and design features of mill tailings (waste) minimize the release of radon and the disturbance of tailings by natural forces (see Glossary)

- developing requirements to ensure cleanup of active and formerly active uranium recovery facilities

- formulating stringent financial requirements to ensure funds are available for decommissioning

- monitoring adherence to requirements for below-grade disposal of mill tailings and liners for tailings impoundments

- monitoring to prevent ground water contamination

- long-term monitoring and oversight of decommissioned facilities

FUEL CYCLE FACILITIES

Special fuel facilities use a process that turns uranium from the ground into fuel for nuclear reactors. This process converts uranium yellowcake into uranium hexafluoride (UF_6), enriches the uranium in the isotope uranium-235, and fabricates ceramic fuel pellets. The NRC licenses and routinely conducts safety, safeguards, and environmental protection inspections at all commercial fuel cycle facilities involved in conversion, enrichment, and fuel fabrication (see Figures 37–40).

Figure 37. Locations of Fuel Cycle Facilities

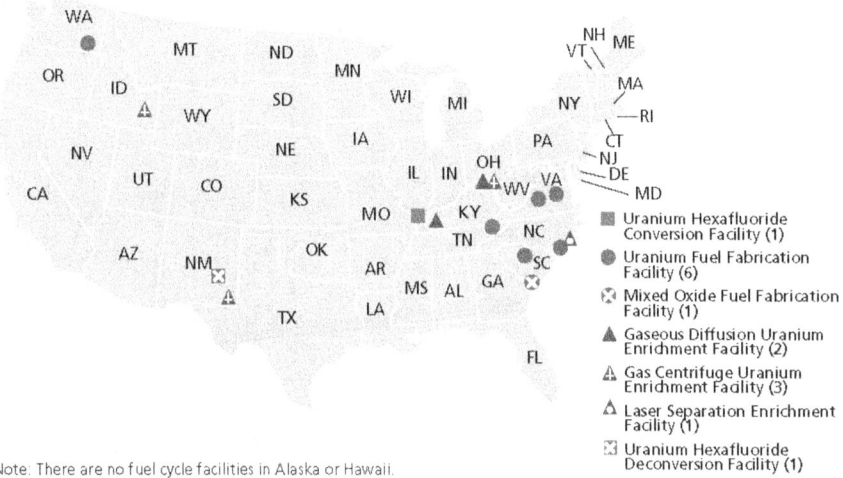

- ◼ Uranium Hexafluoride Conversion Facility (1)
- ⬤ Uranium Fuel Fabrication Facility (6)
- ⊠ Mixed Oxide Fuel Fabrication Facility (1)
- ▲ Gaseous Diffusion Uranium Enrichment Facility (2)
- △ Gas Centrifuge Uranium Enrichment Facility (3)
- △ Laser Separation Enrichment Facility (1)
- ⊠ Uranium Hexafluoride Deconversion Facility (1)

Note: There are no fuel cycle facilities in Alaska or Hawaii.

Table 10. Major U.S. Fuel Cycle Facility Sites

Licensee	Location	Status
Uranium Hexafluoride Conversion Facility		
Honeywell International, Inc.	Metropolis, IL	active
Uranium Fuel Fabrication Facilities		
Global Nuclear Fuels-Americas, LLC	Wilmington, NC	active
Westinghouse Electric Company, LLC Columbia Fuel Fabrication Facility	Columbia, SC	active
Nuclear Fuel Services, Inc.	Erwin, TN	active
AREVA NP, Inc. Mt. Athos Road Facility	Lynchburg, VA	inactive—possession only
B&W Nuclear Operations Group	Lynchburg, VA	active
AREVA NP, Inc.	Richland, WA	active
Mixed Oxide Fuel Fabrication Facility		
Shaw AREVA MOX Services, LLC	Aiken, SC	in construction, operating license under review
Gaseous Diffusion Uranium Enrichment Facilities		
USEC Inc.	Paducah, KY	active
USEC Inc.	Piketon, OH	in cold shutdown*
Gas Centrifuge Uranium Enrichment Facilities		
USEC Inc.	Piketon, OH	in construction
Louisiana Energy Services (URENCO-USA)	Eunice, NM	active**
AREVA Enrichment Services	Idaho Falls, ID	under review
Laser Separation Enrichment Facility		
GE-Hitachi	Wilmington, NC	under review
Uranium Hexafluoride Deconversion Facility		
International Isotopes	Hobbes, NM	under review

* Currently in cold shutdown and in process of decertification and not used for enrichment.

** Partially operating and producing enriched uranium while undergoing further phases of construction.

Note: The NRC regulates nine other facilities that possess significant quantities of special nuclear material (other than reactors) or process source material (other than uranium recovery facilities). Data as of April 2011.

Figure 38. Enrichment Processes

A Gaseo   Diffusion Process

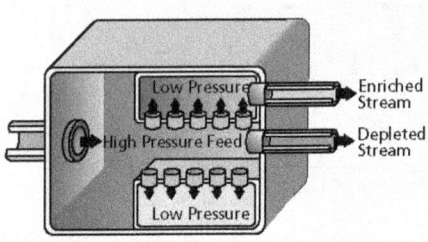

A. *The gaseous diffusion process uses molecular diffusion to separate a gas from a two-gas mixture. The isotopic separation is accomplished by diffusing uranium, which has been combined with fluorine to form UF_6 gas, through a porous membrane (barrier) and using the different molecular velocities of the two isotopes to achieve separation.*

B. *The gas centrifuge process uses a large number of rotating cylinders in series and parallel configurations. Gas is introduced and rotated at high speed, concentrating the component of higher molecular weight toward the outer wall of the cylinder and the component of lower molecular weight toward the center. The enriched and the depleted gases are removed by scoops.*

B. Gas Centrifuge Process

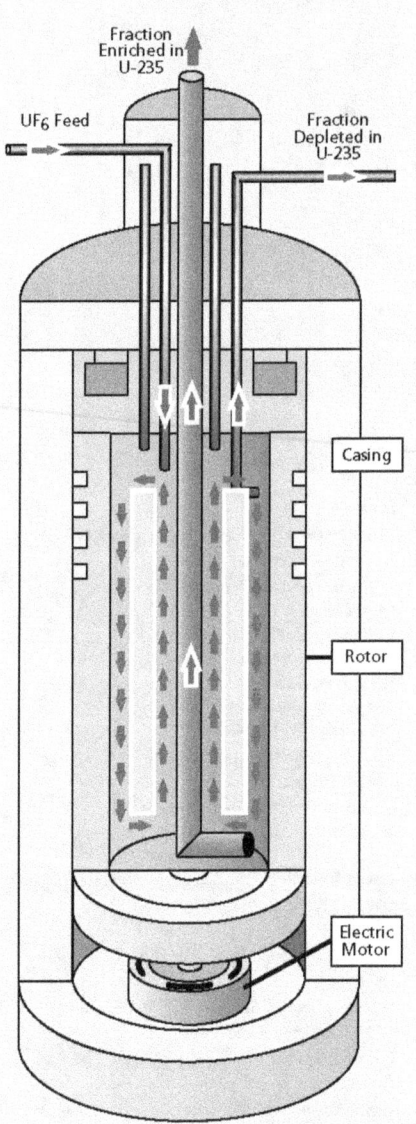

Figure 39. Simplified Fuel Fabrication Process

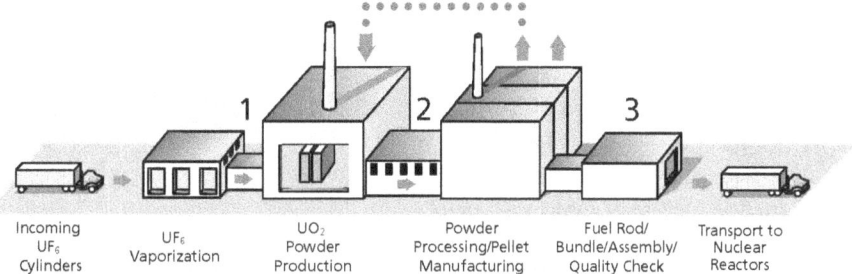

| Incoming UF₆ Cylinders | UF₆ Vaporization | UO₂ Powder Production | Powder Processing/Pellet Manufacturing | Fuel Rod/ Bundle/Assembly/ Quality Check | Transport to Nuclear Reactors |

Fabrication of commercial light-water reactor fuel consists of the following three basic steps:
(1) the chemical conversion of UF₆ to UO₂ powder
(2) a ceramic process that converts UO₂ powder to small ceramic pellets
(3) a mechanical process that loads the fuel pellets into rods and constructs finished fuel assemblies

Figure 40. Fuel Pellets

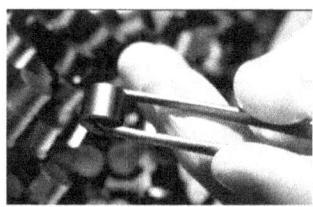

(Left) Small ceramic fuel pellets. (Right) Fuel pellets being assembled into fuel rods.

On average, the NRC completes approximately 85 new licenses, license renewals, license amendments, and safety and safeguards reviews for fuel cycle facilities annually.

Fabrication is the final step in the process used to produce uranium fuel. Fuel fabrication facilities mechanically and chemically process the enriched uranium into nuclear reactor fuel.

Fabrication begins with the conversion of enriched UF_6 gas to a uranium dioxide (UO_2) solid. Nuclear fuel is made to maintain both its chemical and physical properties under the extreme conditions of heat and radiation present inside an operating reactor vessel. After the UF_6 is chemically converted to UO_2, the powder is blended, milled, pressed, and fused into ceramic fuel pellets about the size of a fingertip. The pellets are stacked into tubes about 14 feet (2.6 meters) long made of material called "cladding" (such as zirconium alloys) (see Figure 40). After careful inspection, the resulting fuel rods are bundled into fuel assemblies for use in reactors. The assemblies are washed, inspected, and stored in a special rack until ready for shipment to a nuclear power plant site. The NRC inspects this operation at every step of the process.

NUCLEAR MATERIALS

Domestic Safeguards Program

The NRC's domestic safeguards program for fuel cycle facilities and transportation is aimed at ensuring that special nuclear material (such as plutonium or enriched uranium) is not stolen for possible malevolent uses. The program also works to ensure that such material does not pose an unreasonable risk to the public from sabotage or terrorism.

The NRC verifies through licensing and inspection activities that licensees apply safeguards to protect special nuclear material. Additionally, the NRC and DOE developed the Nuclear Materials Management and Safeguards System (NMMSS) to track transfers and inventories of special nuclear material, source material from abroad, and other material.

The NRC has issued licenses to approximately 180 facilities authorizing them to possess special nuclear material in quantities ranging from a single kilogram to multiple tons. These licensees verify and document their inventories in the NMMSS database. The NRC or State governments license several hundred additional sites that possess special nuclear material in smaller quantities (typically ranging from 1 gram to tens of grams).

Licensees that possess small amounts of special nuclear material are now required to confirm their inventory annually in the NMMSS database. Previously, those licensees reported transfers of material but not annual inventories.

RADIOACTIVE WASTE

Top: Dry cask storage of spent nuclear fuel.

Middle: Spent fuel pool at a research and test reactor. (Photo courtesy: University of Wisconsin-Madison)

Bottom: NRC inspectors examine a container to determine if it meets NRC standards.

LOW-LEVEL RADIOACTIVE WASTE DISPOSAL

Low-level radioactive waste (LLW) includes items that have become contaminated with radioactive material or have become radioactive through exposure to neutron radiation. This waste typically consists of contaminated protective shoe covers and clothing, wiping rags, mops, filters, reactor water treatment residues, equipment and tools, medical tubes, swabs, injection needles, syringes, and laboratory animal carcasses and tissue.

The radioactivity can range from just-above-background levels found in nature to very high levels from the parts inside the reactor vessel in a nuclear power plant. Licensees store some lower level radioactive waste onsite until it has decayed and lost its radioactivity. Then it can be disposed of as ordinary trash. Waste that does not decay fairly quickly is stored until amounts are large enough for shipment to an LLW disposal site

Figure 41. Low-Level Waste Disposal

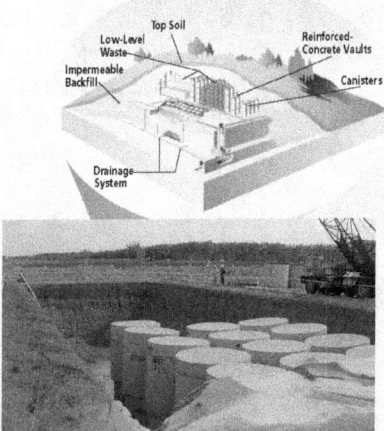

This LLW disposal site accepts waste from the compact States.

in containers approved by DOT and the NRC.

Commercial LLW is disposed of in facilities licensed by either the NRC or Agreement States in accordance with health and safety requirements. The facilities are designed, constructed, and operated to meet safety standards. The operator of the facility also extensively characterizes the site on which the facility is located and analyzes how the facility will perform in the future. Current LLW disposal uses shallow land disposal sites with or without concrete vaults (see Figure 41).

The NRC has developed a classification system for LLW based on its potential hazards. It has specified disposal and waste requirements for each of the three classes of waste—Class A, B, and C—that are acceptable for disposal in near-surface facilities. These classes have progressively higher levels of concentrations of radioactive material, with A having the lowest and C having the highest level. Class A waste accounts for approximately 96 percent of the total volume of LLW. Determination of the classification of waste is a complex process. A fourth class of LLW, greater than Class C, is not generally acceptable for near-surface, shallow-depth disposal. By law, DOE is responsible for disposal of greater than Class C waste under an NRC license.

The volume and radioactivity of waste vary from year to year based on the types and quantities of waste shipped each year. Waste volumes currently include several million cubic feet each year from reactor facilities undergoing decommissioning and from cleanup of contaminated sites.

The Low-Level Radioactive Waste Policy Amendments Act of 1985 gave the States responsibility for the disposal of LLW. The Act authorized States to do the following:

- Form regional compacts, with each compact to provide for LLW disposal site access (see Table 11).
- Manage LLW import to, and export from, a compact.

Exclude waste generated outside a compact.

The States have licensed four active LLW disposal facilities:

- **Barnwell**, located in Barnwell, SC— Previously, Barnwell accepted waste from all U.S. generators. As of July 2008, Barnwell accepts waste from the Atlantic Compact States (Connecticut, New Jersey, and South Carolina). The State of South Carolina licenses Barnwell to receive Classes A, B, and C of LLW.
- **EnergySolutions**, located in Clive, UT—EnergySolutions accepts waste from all regions of the United States. Utah licenses EnergySolutions for Class A waste only.

- **Hanford**, located in Hanford, WA—Hanford accepts waste from the Northwest and Rocky Mountain Compacts. The State of Washington licenses Hanford to receive Classes A, B, and C of LLW.
- **Waste Control Specialist** (WCS), located in Andrews, TX—The State of Texas licensed WCS in 2009 to receive Classes A, B, and C of LLW from the Texas Compact, which consists of Texas and Vermont. WCS is expected to begin receiving LLW in late 2011.

Closed LLW disposal facilities licensed by the NRC or Agreement States include the following:

- Beatty, NV, closed 1993
- Sheffield, IL, closed 1978
- Maxey Flats, KY, closed 1977
- West Valley, NY, closed 1975

Table 11. U.S. Low-Level Radioactive Waste Compacts

Appalachian

Delaware
Maryland
Pennsylvania
West Virginia

Atlantic

Connecticut
New Jersey
South Carolina*

Central

Arkansas
Kansas
Louisiana
Oklahoma

Central Midwest

Illinois
Kentucky

Midwest

Indiana
Iowa
Minnesota
Missouri
Ohio
Wisconsin

Northwest

Alaska
Hawaii
Idaho
Montana
Oregon
Utah*
Washington*
Wyoming

Rocky Mountain

Colorado
Nevada
New Mexico
(Northwest accepts Rocky Mountain waste as agreed between compacts)

Southeast

Alabama
Florida
Georgia
Mississippi
Tennessee
Virginia

Southwestern

Arizona
California
North Dakota
South Dakota

Texas

Texas
Vermont

Unaffiliated

District of Columbia
Maine
Massachusetts
Michigan
Nebraska
New Hampshire
New York
North Carolina
Puerto Rico
Rhode Island

Note: Data as of June 2011.
*Site of an active LLW disposal facility.

HIGH-LEVEL RADIOACTIVE WASTE MANAGEMENT

Spent Nuclear Fuel Storage

Commercial spent nuclear fuel, although highly radioactive, is stored safely and securely in 35 States (see Figure 42). This includes 31 States with operating nuclear power reactors, where spent fuel is safely stored onsite in spent fuel pools and in some dry casks. The remaining four States—Colorado, Idaho, Maine, and Oregon—do not have operating power reactors but are safely storing spent fuel at storage facilities. Waste can be stored safely in pools or casks for 100 years or more.

As of January 2011, the amount of commercial spent fuel in safe storage at commercial nuclear power plants was an estimated 63,000 metric tons. The amount of spent fuel in storage at individual commercial nuclear power plants is expected to increase at a rate of approximately 2,000 metric tons per year. The NRC licenses and regulates the storage of spent fuel, both at commercial nuclear power plants and at storage facilities located away from reactors.

Most reactor facilities were not designed to store the full amount of spent fuel that the reactor would generate during its operational life. Facilities originally planned to store spent fuel temporarily in deep pools of continuously circulating water that cools the spent fuel assemblies and provides shielding from radiation. After a few years, the facilities expected

Figure 42. Storage of Commercial Spent Fuel by State through 2011

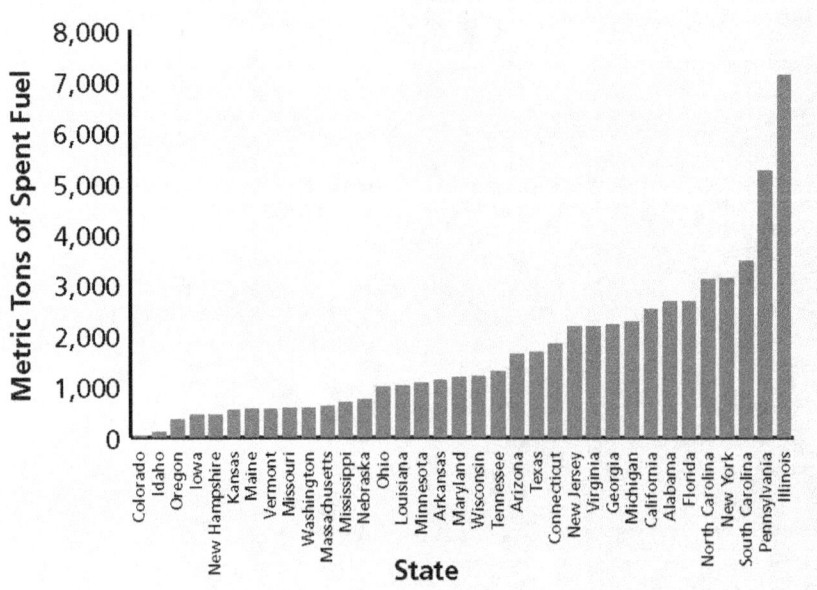

Note: Idaho is holding used fuel from Three Mile Island, Unit 2. Data are rounded up to the nearest 10 tons
Source: ACI Nuclear Energy Solutions and U S Department of Energy (updated May 2011)

to send the spent fuel to a recycling plant. However, the Federal Government declared a moratorium on recycling spent fuel in 1977. Although the ban was later lifted, recycling has not been pursued. To cope with the spent fuel they were generating, facilities expanded their storage capacity by using high-density storage racks in their spent fuel pools (see Figure 43). However, spent fuel pools are not a permanent storage solution.

To provide supplemental storage, a portion of spent fuel inventories is stored in dry casks on site. These facilities are called independent spent fuel storage installations (ISFSIs) and are licensed by the NRC. These large casks are typically made of leak-tight, welded, and bolted steel and concrete surrounded by another layer of steel or concrete. The spent fuel sits in the center of the nested canisters in an inert gas. Dry cask storage shields people and the environment from radiation and keeps the spent fuel inside dry and nonreactive (see Figure 44).

Currently, there are 63 licensed ISFSIs in the United States. As of 2011, NRC-licensed ISFSIs were storing spent fuel in over 1,220 loaded dry casks (see Figure 45).

The NRC authorizes storage of spent fuel at an ISFSI under two licensing options:

1. site-specific licensing

2. general licensing

Site-specific licenses granted by the NRC after a safety review contain technical requirements and operating conditions for the ISFSI and specify what the licensee is authorized to store at the site. The initial and renewal license terms for an ISFSI are not to exceed 40 years from the date of issuance.

A general license from the NRC authorizes a licensee who operates a nuclear power reactor to store spent fuel onsite in dry storage casks. Under the general license, the authority to use a storage cask is tied to the cask's Certificate of Compliance (CoC) term. A CoC is issued to the cask vendor through rulemaking. Several dry storage cask designs have received certificates. Initial and renewed CoCs are issued for terms not to exceed 40 years.

See Appendix H for a list of dry spent fuel storage systems that are approved for use with a general license. See Appendix I for a list of dry spent fuel storage licensees.

No more than 30 days before the certificate expiration date, the cask vendor may apply for renewal. If the cask vendor does not apply for renewal, a general licensee may apply for renewal.

Before using the cask, general licensees must certify that the cask meets the conditions in the certificate, that the concrete pads under the casks can adequately support the loads, and that the levels of radiation from the casks meet NRC standards. Specific license and CoC renewal applications must include an analysis that considers the effects of aging on structures, systems, and components of safety for the requested renewal term.

RADIOACTIVE WASTE

Figure 43. Spent Fuel Generation and Storage after Use

1 *A nuclear reactor is powered by enriched uranium-235 fuel. Fission (splitting of atoms) generates heat, which produces steam that turns turbines to produce electricity. A reactor rated at several hundred megawatts may contain 100 or more tons of fuel in the form of bullet-sized pellets loaded into long metal rods that are bundled together into fuel assemblies. PWRs contain between 150 and 200 fuel assemblies. BWRs contain between 370 and 800 fuel assemblies.*

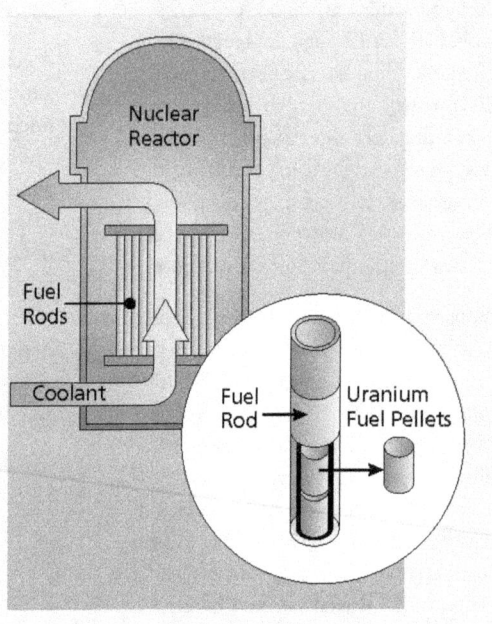

2 *After about 6 years, spent fuel assemblies—typically 14 feet (4.3 meters) long and containing nearly 200 fuel rods for PWRs and 80–100 fuel rods for BWRs—are removed from the reactor and allowed to cool in storage pools for a few years. At this point, the 900-pound (409-kilogram) assemblies contain only about one-fifth the original amount of uranium-235.*

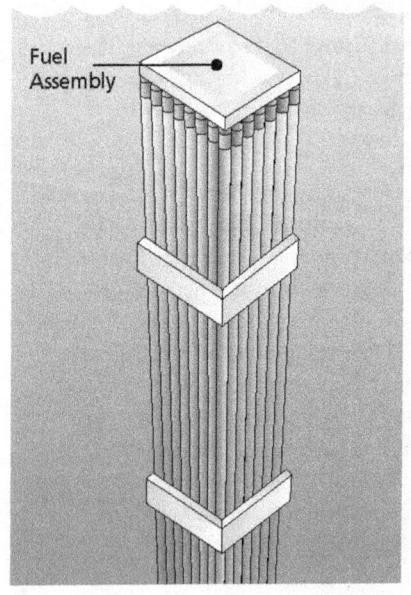

3 Commercial light-water nuclear reactors store spent radioactive fuel in a steel-lined,
   seismically designed concrete pool under about 40 feet (12.2 meters) of water that
provides shielding from radiation. Water pumps supply continuously flowing water to cool
the spent fuel. Extra water for the pool is provided by other pumps that can be powered
from an onsite emergency diesel generator. Support features, such as water-level monitors
and radiation detectors, are also in the pool. Spent fuel is stored in the pool until it can be
transferred to dry casks on site (as shown in Figure 44) or transported off site to a high-level
radioactive waste disposal site.

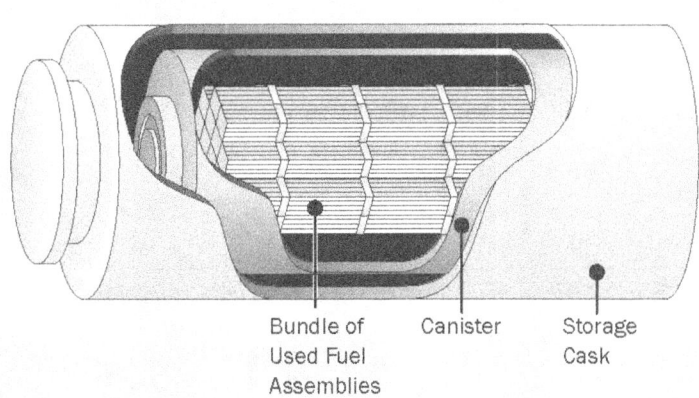

Bundle of   Canister   Storage
Used Fuel   Cask
Assemblies

Source: DOE and the Nuclear Energy Institute

Figure 44. Dry Storage of Spent Nuclear Fuel

At some nuclear reactors across the country, spent fuel is kept onsite, typically above ground, in systems basically similar to the ones shown here.

1 *Once the spent fuel has cooled, it is loaded into special canisters that are designed to hold nuclear fuel assemblies. Water and air are removed. The canister is filled with inert gas, welded shut, and rigorously tested for leaks. It is then placed in a cask for storage or transportation. The NRC has approved the storage of up to 40 PWR assemblies and up to 68 BWR assemblies in each canister. The dry casks are then loaded onto concrete pads.*

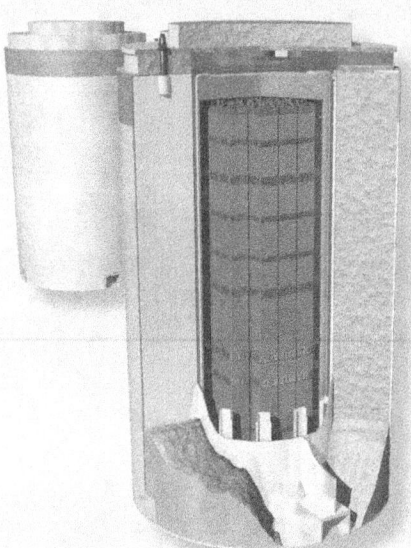

2 *The canisters can also be stored in aboveground concrete bunkers, each of which is about the size of a one-car garage.*

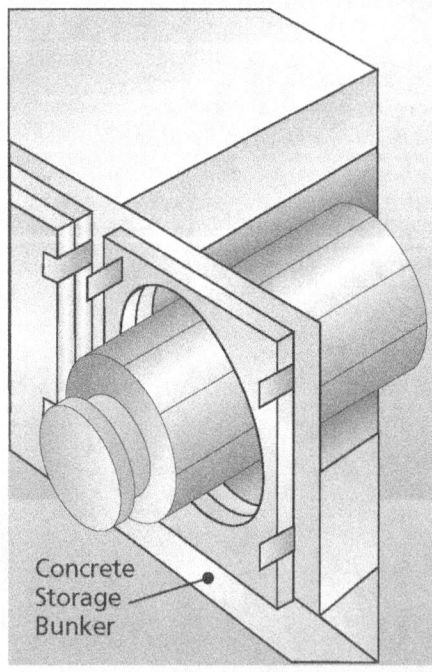

Concrete Storage Bunker

Figure 45. Licensed/Operating Independent Spent Fuel Storage Installations by State

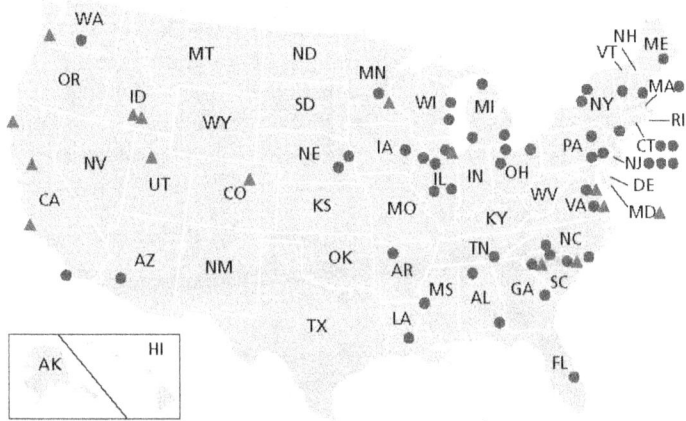

33 States have at least one ISFSI

▲ Site-Specific License (15)
● General License (48)

ALABAMA
● Browns Ferry
● Farley

ARIZONA
● Palo Verde

ARKANSAS
● Arkansas Nuclear

CALIFORNIA
▲ Diablo Canyon
▲ Rancho Seco
● San Onofre
▲ Humboldt Bay

COLORADO
▲ Fort St. Vrain

CONNECTICUT
● Haddam Neck
● Millstone

FLORIDA
● St. Lucie

GEORGIA
● Hatch

IDAHO
▲ DOE: TMI-2 (Fuel Debris)
▲ Idaho Spent Fuel Facility

ILLINOIS
● Byron
▲ GE Morris (Wet)
● Dresden
● La Salle
● Quad Cities

IOWA
● Duane Arnold

LOUISIANA
● River Bend

MAINE
● Maine Yankee

MARYLAND
▲ Calvert Cliffs

MASSACHUSETTS
● Yankee Rowe

MICHIGAN
● Big Rock Point
● Fermi
● Palisades

MINNESOTA
● Monticello
▲ Prairie Island

MISSISSIPPI
● Grand Gulf

NEBRASKA
● Cooper
● Ft. Calhoun

NEW HAMPSHIRE
● Seabrook

NEW JERSEY
● Hope Creek
● Salem
● Oyster Creek

NEW YORK
● Indian Point
● FitzPatrick
● Ginna

NORTH CAROLINA
● Brunswick
● McGuire

OHIO
● Davis-Besse
● Perry

OREGON
▲ Trojan

PENNSYLVANIA
● Limerick
● Susquehanna
● Peach Bottom

SOUTH CAROLINA
● ▲ Oconee
● ▲ Robinson
● Catawba

TENNESSEE
● Sequoyah

UTAH
▲ Private Fuel Storage

VERMONT
● Vermont Yankee

VIRGINIA
● ▲ Surry
● ▲ North Anna

WASHINGTON
● Columbia

WISCONSIN
● Point Beach
● Kewaunee

Note: Data are current as of June 2011. NRC-abbreviated unit names used.

Public Involvement

The public can participate in decisions about spent fuel storage, as it can in many licensing and rulemaking decisions. The Atomic Energy Act of 1954, as amended, and the NRC's own regulations provide the opportunity for public hearings for site-specific licensing actions and allow for public comments on certificate rulemakings. Interested members of the public may also file petitions for rulemaking.

Additional information on ISFSIs is available on the NRC Web site (see the Web Link Index).

Spent Nuclear Fuel Disposal

The current U.S. policy governing permanent disposal of high-level radioactive waste is defined by the Nuclear Waste Policy Act of 1982, as amended, and the Energy Policy Act of 1992. These acts specify that high-level radioactive waste will be disposed of underground in a deep geologic repository.

DOE submitted its license application to the NRC on June 3, 2008, for Yucca Mountain in Nevada. The NRC formally accepted it for review

in September 2008 and began the detailed technical review and associated adjudicatory activities. In 2009, President Barack Obama announced that the administration would terminate the Yucca Mountain program while developing a disposal alternative. The NRC will complete an orderly closeout of the Yucca Mountain project.

On January 29, 2010, the President created the Blue Ribbon Commission on America's Nuclear Future to reassess the national policy on high-level waste disposal. The task of the Blue Ribbon Commission is to "conduct a comprehensive review of policies for managing the back end of the nuclear fuel cycle." In light of these developments, the NRC began reassessing its management of spent fuel regulations to position the agency to adapt quickly to changes in national policy. The three key areas in this effort are the nuclear fuel cycle, spent fuel storage and transportation, and high-level waste disposal.

Recycling

In the United States, spent nuclear fuel is stored safely and securely either at a nuclear power plant or at a storage facility away from a plant. Some

Figure 46. Ensuring Safe Spent Fuel Shipping Containers

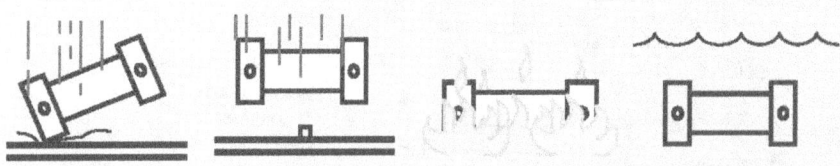

The impact (free drop and puncture), fire, and water-immersion tests are considered in sequence to determine their cumulative effects on a given package.

countries reprocess their spent nuclear fuel to recover fissile material and use it to generate more energy. Although the NRC has not received an application for a reprocessing facility, the agency has completed an initial analysis of the existing regulatory framework in preparation for such an application. The NRC is developing the technical basis for a possible revision of the regulations to ensure that a potential commercial reprocessing facility can be licensed efficiently and effectively and operate safely.

TRANSPORTATION

The NRC is also involved in the transportation of spent nuclear fuel. It establishes safety criteria for spent fuel shipping casks and certifies cask designs. Casks are designed to meet the following safety criteria under both normal and accident conditions:

- Prevent the loss or dispersion of radioactive contents.

- Provide shielding and heat dissipation.

- Prevent nuclear criticality (a self-sustaining nuclear chain reaction).

Spent fuel shipping casks must be designed to survive a sequence of tests, including a 9-meter (30-foot) drop onto an unyielding surface, a puncture test, and a fully engulfing fire at 1,475 degrees Fahrenheit (802 degrees Celsius) for 30 minutes. This very severe test sequence, akin to the cask striking a concrete pillar along a highway at a high speed and being engulfed in a very severe and long-lasting fire, simulates conditions more severe than 99 percent of vehicle accidents (see Figure 46).

Empty storage transport container on a semi tractor-trailer rig.

Principal Licensing and Inspection Activities

The NRC regulates spent fuel transportation through a combination of safety and security requirements, certification of transportation casks, inspections, and a system of monitoring to ensure that requirements are being met.

Specifically, each year, the NRC does the following:

- Conducts about 1,000 transportation safety inspections of fuel, reactor, and materials licensees.

- Reviews, evaluates, and certifies approximately 80 new, renewal, or amended transport package design applications.

- Inspects about 20 dry storage and transport package licensees.

- Reviews and evaluates approximately 150 license applications for the import or export of nuclear materials.

Additional information on materials transportation is available on the NRC Web site (see the Web Link Index).

DECOMMISSIONING

Decommissioning is the safe removal of a nuclear facility from service and the reduction of residual radioactivity to a level that permits release of the property and termination of the license. The NRC rules establish site-release criteria and provide for unrestricted and, under certain conditions, restricted release of a site.

The NRC regulates the decontamination and decommissioning of materials and fuel cycle facilities, nuclear power plants, research and test reactors, and uranium recovery facilities, with the ultimate goal of license termination. The NRC terminates approximately 200 materials licenses each year. Most of these license terminations are routine, and the sites require little, if any, remediation to meet the NRC's release criteria for unrestricted access. The decommissioning program focuses on the termination of licenses that are not routine because the sites involve more complex decommissioning activities (see Figure 47).

As of early 2011, the following facilities were undergoing decommissioning under NRC jurisdiction:

- 13 nuclear power early demonstration reactors

- 12 research and test reactors

- 20 complex decommissioning materials facilities (see Table 12)

- 1 fuel cycle facility

- 11 uranium recovery facilities

  See Appendices B and F for lists of permanently shut down nuclear reactors and nuclear power, research, and test reactors undergoing decommissioning.

The "Status of the Decommissioning Program 2010 Annual Report" provides additional information on the decommissioning programs of the NRC and Agreement States. More information is on the NRC Web site (see the Web Link Index).

As part of the decommissioning process, the cooling tower of a nuclear power plant is imploded.

Figure 47. Locations of NRC-Regulated Complex Material Sites Undergoing Decommissioning

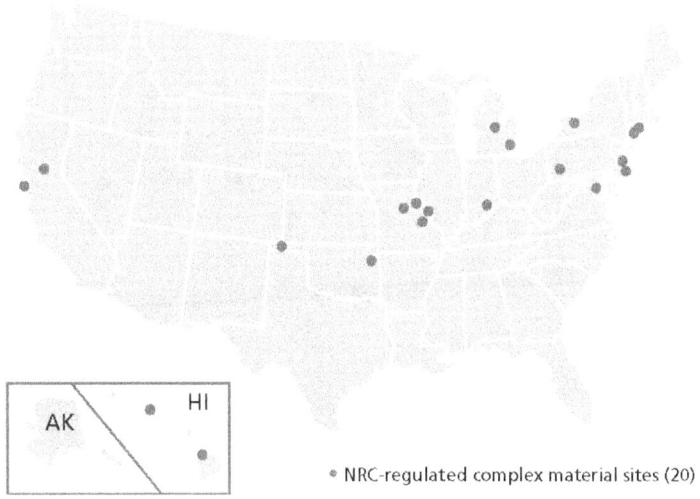

• NRC-regulated complex material sites (20)

Table 12. NRC-Regulated Complex Material Sites Undergoing Decommissioning

Company	Location
AAR Manufacturing, Inc. (Brooks & Perkins)	Livonia, MI
ABB, Inc.	Windsor, CT
Analytical Bio-Chemistry Laboratories	Columbia, MO
Army, Department of, Jefferson Proving Ground	Madison, IN
Babcock & Wilcox SLDA	Vandergrift, PA
Beltsville Agricultural Research Center	Beltsville, MD
FMRI	Muskogee, OK
Hunter's Point Naval Shipyard	San Francisco, CA
Kerr-McGee	Cimarron, OK
Mallinckrodt Chemical, Inc.	St. Louis, MO
McClellan Air Force Base	Sacramento, CA
NWI Breckenridge	Breckenridge, MI
Pohakuloa Training Area	Kawaihe Harbor, HI
Shieldalloy Metallurgical Corp.	Newfield, NJ
Schofield Army Barracks	Wahiawa, HI
Sigma Aldrich	Maryland Heights, MO
Stepan Chemical Corporation	Maywood, NJ
UNC Naval Products	New Haven, CT
West Valley Demonstration Project	West Valley, NY
Westinghouse Electric Corporation—Hematite	Festus, MO

Note: Data as of July 2011.

SECURITY AND EMERGENCY PREPAREDNESS

Top: Nuclear power plant security officers don special equipment for a mock attack drill.

Middle: The Commissioners listen as NRC Executive Director for Operations Bill Borchardt gives a briefing on the agency's response to the recent nuclear events in Japan.

Bottom: The NRC Headquarters Operations Center during an emergency preparedness exercise.

OVERVIEW

Nuclear security is a high priority for the NRC. For the past several decades, effective NRC regulation and strong partnerships with a variety of Federal, State, Tribal, and local authorities have ensured effective implementation of security programs at nuclear power plants across the country. In fact, nuclear power plants are likely the best protected private sector facilities in the United States. However, given today's threat environment, the agency recognizes the need for continued vigilance and high levels of security.

In recent years, the NRC has made many enhancements to bolster the security of the Nation's nuclear facilities and radioactive materials. Because nuclear power plants are inherently robust structures, these additional security upgrades largely focus on the following improvements:

- well-trained and armed security officers

- high-tech equipment and physical barriers

- greater standoff distances for vehicle checks

- intrusion detection and surveillance systems

- tested emergency preparedness and response plans

- restrictive site access control, including background checks and fingerprinting of workers

Additional layers of security are provided by coordinating and sharing threat information among DHS, the U.S. Federal Bureau of Investigation, intelligence agencies, the U.S. Department of Defense, and local law enforcement.

FACILITY SECURITY

Nuclear power plants and Category I fuel facilities must be able to defend successfully against a set of hypothetical threats that the agency calls the design-basis threat (DBT). This includes threats that challenge a plant's physical security, personnel security, and cyber security. The NRC does not make details of the DBT public because of security concerns. However, the agency continuously evaluates this set of hypothetical threats against real-world intelligence to ensure that the DBT remains current.

To test the adequacy of a nuclear power plant's defenses against the DBT, the NRC conducts rigorous "force-on-force" inspections. During these inspections, exercises are conducted in which a highly trained mock adversary force "attacks" a nuclear facility. Beginning in 2004, the NRC began conducting more challenging and realistic force-on-force exercises that also occur more frequently.

To ensure that facilities meet their security requirements, the NRC inspects nuclear power plants and fuel fabrication facilities, spending about 8,000 hours a year scrutinizing security (excluding force-on-force inspections). Publicly available portions of security-related inspection reports can be found on the NRC Web site (see the Web Link Index).

A well-trained and armed security officer at a nuclear power plant facility.

CYBER SECURITY

Nuclear facilities use digital and analog systems to monitor, control, and run various types of equipment and to obtain and store vital information. Protecting these systems and the information they contain from sabotage or malicious use is called "cyber security." All nuclear power plants licensed by the NRC must have a cyber security program. A new cyber security rule, issued in 2009, requires each nuclear power facility to submit a cyber security plan and implementation timeline for NRC approval. Once the licensee has fully implemented its program, the NRC will conduct a comprehensive inspection on site.

The NRC has formed a cyber security team that includes technology and threat experts who constantly evaluate and identify emerging cyber-related issues that could affect plant systems. This team makes recommendations to other NRC offices and programs on cyber security issues.

MATERIALS SECURITY

The security of radioactive materials is important for a number of reasons. For example, terrorists could use radioactive materials to make a radiological dispersal device such as a dirty bomb. The NRC works with its Agreement States, other Federal agencies, IAEA, and licensees to protect radioactive material from theft or diversion. The agency has made improvements and upgrades to the joint NRC-DOE database that tracks the movement and location of certain forms and quantities of special nuclear material. In early 2009, the NRC deployed its new NSTS, designed to track the most risk-sensitive sources on a continuous basis. Other improvements allow U.S. Customs and Border Protection agents to promptly validate whether radioactive materials coming into the United States are properly licensed by the NRC.

SECURITY AND EMER-GENCY PREPAREDNESS

EMERGENCY PREPAREDNESS

Well-developed and practical emergency preparedness plans ensure that a nuclear power plant operator can protect public health and safety in the unlikely event of an emergency.

The NRC staff participates in emergency preparedness exercises, some of which include security- and terrorism-based scenarios. To form a coordinated system of emergency preparedness and response, as part of these exercises, the NRC works with licensees; Federal agencies; State, Tribal, and local officials; and first responders. This system includes public information, preparations for evacuation, instructions for sheltering, and other actions to protect the residents near nuclear power plants in the event of a serious incident.

As a condition of their license, operators of nuclear facilities develop and maintain effective emergency plans and procedures. The NRC inspects licensees to ensure that they are prepared to deal with emergencies. In addition, the agency monitors performance indicators related to emergency preparedness. (see Figure 48).

The NRC assesses the ability of nuclear power plant operators to respond to emergencies. For nuclear power plants, operators are required to conduct full-scale exercises with the NRC, the Federal Emergency Management Agency (FEMA), and State and local officials at least once every 2 years.

Figure 48. Industry Performance Indicators: Annual Industry Percentages, FYs 2001–2010—for 104 Plants

The percentage of ANS sirens that successfully operated during periodic tests in the previous year. The result is an indicator of the reliability of the ANS to alert the public in an emergency.

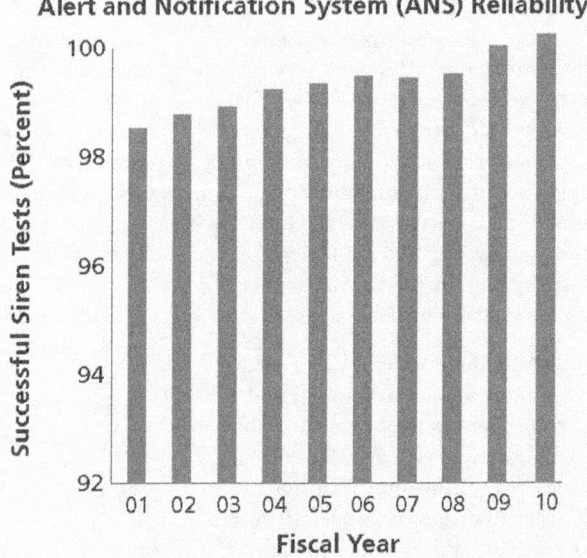

Alert and Notification System (ANS) Reliability

Drill/Exercise Performance

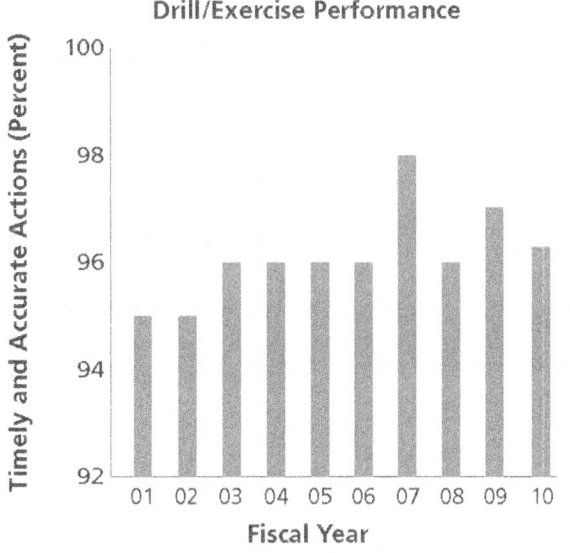

The percentage of timely and accurate actions taken by plant personnel (emergency classifications, protective action recommendations, and notification to offsite authorities) in drills and actual events during the previous 2 years.

Emergency Response Organization (ERO) Drill Participation

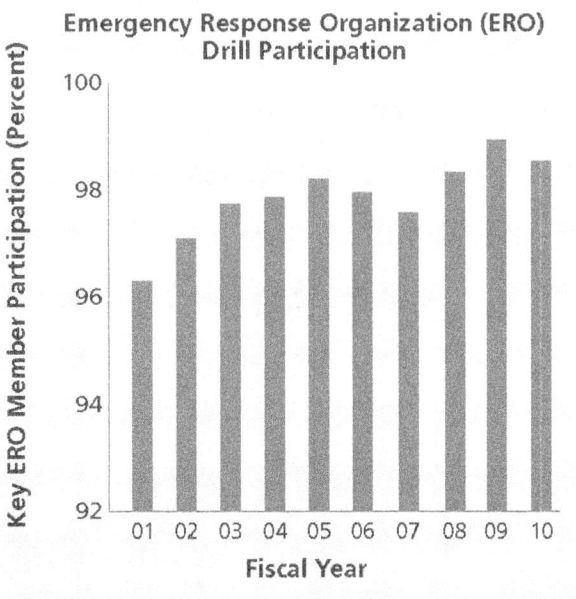

The percentage of participation by key plant personnel in drills or actual events in the previous 2 years, indicating proficiency and readiness to respond to emergencies.

SECURITY AND EMER-
GENCY PREPAREDNESS

These exercises test and maintain the skills of the emergency responders and identify areas that need addressing. The NRC and FEMA evaluate these exercises. Between these 2-year exercises, nuclear power plant operators self-test their emergency plans in drills that NRC inspectors evaluate.

Emergency Planning Zones

Although emergency planning zones (EPZs) are meant to be expanded, as necessary, for planning purposes, the NRC defines two zones around each nuclear power plant. The exact size and configuration of the zones vary from plant to plant based on local emergency response needs and capabilities, population, land characteristics, access routes, and jurisdictional boundaries (see Figure 49 for a typical EPZ around a nuclear plant). The two types of EPZs are as follows:

- The plume exposure pathway EPZ extends about 10 miles in radius around a plant. Its primary concern is the exposure of the public to, and the inhalation of, airborne radioactive contamination. Research has shown that the most significant impacts of an accident would be expected in the immediate vicinity of a plant, and any initial protective actions such as evacuations or sheltering in place should be focused there.

- The ingestion pathway EPZ extends about 50 miles in radius around a plant. Its primary concern is the ingestion of food and liquid that is contaminated by radioactivity.

During an actual radiological event, the NRC would perform dose calculations using radiation dose projection models that analyze release paths from power reactors. The dose calculations would also

Figure 49. Emergency Planning Zones

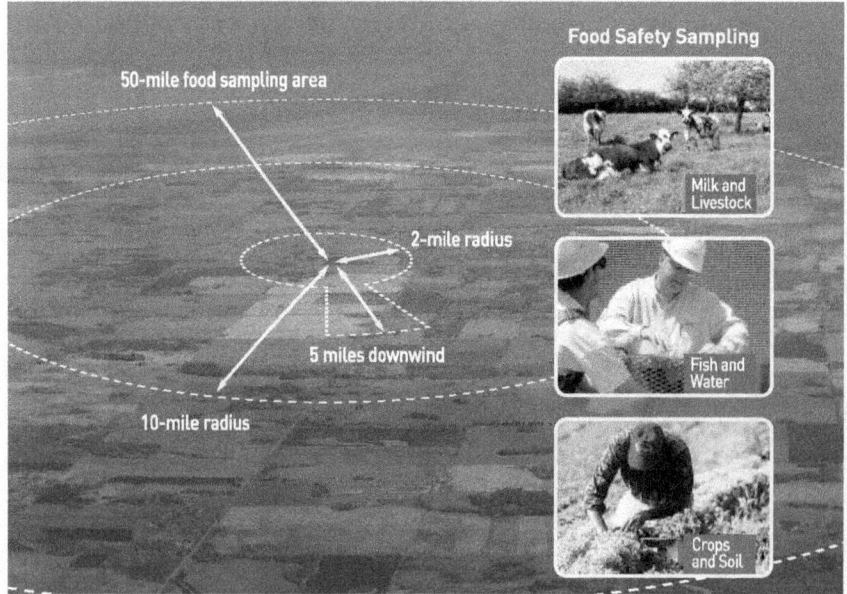

take into account weather conditions to project radiation doses. The NRC would confer with appropriate State and county governments on its assessment results. Plant personnel would also provide assessments. State and local officials in communities within the EPZ have detailed plans to protect public health and safety in the event of a radiological release. These officials make the protective action decisions based on the assessments that they have received.

Evacuation, Sheltering, and the Use of Potassium Iodide

Protective actions considered for a radiological emergency include evacuation, sheltering, and, as a supplement to these, the prophylactic use of potassium iodide (KI) to protect the thyroid from absorbing radioactive iodine. Under most conditions, evacuation may be preferred to remove the public from further exposure to radioactive material. However, under some conditions, people may be instructed to take shelter in their homes, schools, or office buildings. Depending on the type of structure, sheltering can significantly reduce a person's dose compared to the dose received if he or she remained outside. In certain situations, KI is used as a supplement to sheltering.

Evacuation does not always call for the complete evacuation of the 10-mile zone around a nuclear power plant. In most cases, the release of radioactive material from a plant during a major incident would move with the wind, not in all directions surrounding the plant. The release would also expand and become less concentrated as it travels away from a plant. Therefore, evacuations should be mapped to anticipate the path of the release.

Sheltering is a protective action that keeps people indoors to reduce exposure to radioactive material. It may be appropriate to shelter when the release of radioactive material is known to be short term or is controlled by the nuclear power plant operator.

Additional information on emergency preparedness is available on the NRC Web site (see the Web Link Index).

INCIDENT RESPONSE

Sharing information quickly among the NRC, other Federal agencies, and the nuclear industry is critical to responding promptly to any incident. The NRC staff supports several important Federal incident response centers that coordinate assessments of event-related information. The NRC Headquarters Operations Center, located in the agency's headquarters in Rockville, MD, is staffed around the clock to disseminate information and coordinate response activities. To ensure the timely distribution of threat information, the NRC reviews intelligence reports and assesses suspicious activity.

The NRC works within the National Response Framework to respond to events. The framework guides the Nation in how to respond to complex events that may involve a variety of agencies and hazards.

Under this framework, the NRC retains its independent authority and ability to respond to emergencies that involve

The NRC expanded 24-hour coverage of its operations center following the March 2011 earthquake and tsunami event in Japan. Staff provided support to overseas counterparts and examined available information to understand implications for the United States.

NRC-licensed facilities or materials. The NRC coordinates the Federal technical response to an incident that involves one of its licensees.

The NRC may request DHS support in responding to an emergency at an NRC-licensed facility or involving NRC-licensed materials. DHS may lead and manage the overall Federal response to an event, according to Homeland Security Presidential Directive 5, "Management of Domestic Incidents," dated February 28, 2003. In this case, the NRC would provide technical expertise and help share information among the various organizations and licensees.

In response to an incident involving possible releases of radioactive

materials, the NRC activates its incident response program at its Headquarters Operations Center and one of its four Regional Incident Response Centers. Teams of specialists assemble at these Centers to evaluate event information and independently assess the potential impact on public health and safety. The NRC staff provides expert consultation, support, and assistance to State and local public safety officials and keeps the public informed of agency actions. Scientists and engineers at these Centers analyze the event and evaluate possible recovery strategies. Meanwhile, other NRC experts evaluate the effectiveness of protective actions the licensee has recommended that State and local officials implement. If needed, the NRC will dispatch a team of technical experts from the responsible regional office to the site of the incident. Augmenting the NRC's resident inspectors, who work at the plant, the team serves as the agency's onsite eyes and ears, allowing a firsthand assessment and face-to-face communications with all participants. The Headquarters Operations Center continues to provide around-the-clock Federal communications, logistical support, and technical analysis throughout the response.

Emergency Classifications

Based on NRC regulations, the emergency classifications are the sets of plant conditions that indicate various levels of risk to the public and that might require response by an offsite emergency response organization to protect citizens near the site.

Both nuclear power plants and research and test reactors use the following four emergency classifications:

- *Notification of Unusual Event*— Events that indicate potential degradation in the level of safety of the plant are in progress or have occurred. No release of radioactive material requiring offsite response or monitoring is expected unless further degradation occurs.

- *Alert*—Events that involve an actual or potential substantial degradation in the level of plant safety are in progress or have occurred. Any releases of radioactive material are expected to be limited to a small fraction of the limits set forth by the U.S. Environmental Protection Agency (EPA).

- *Site Area Emergency*—Events that may result in actual or likely major failures of plant functions needed to protect the public are in progress or have occurred. Any releases of radioactive material are not expected to exceed the limits set forth by EPA except near the site boundary.

- *General Emergency*—Events that involve actual or imminent substantial core damage or melting of reactor fuel with the potential for loss of containment integrity are in progress or have occurred. Radioactive releases can be expected to exceed the limits set forth by EPA for more than the immediate site area.

Nuclear materials and fuel cycle facility licensees use the following emergency classifications:

- *Alert*—Events that could lead to a release of radioactive materials are in progress or have occurred. The release is not expected to require a response by an offsite response organization to protect citizens near the site.

- *Site Area Emergency*—Events that could lead to a significant release of radioactive materials are in progress or have occurred. The release could require a response by offsite response organizations to protect citizens near the site.

APPENDICES

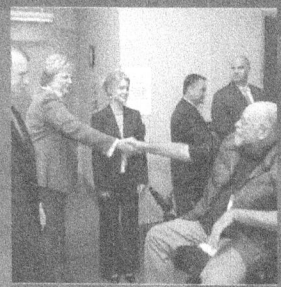

Top: NRC staff at the agency's Annual Health Fair.

Center: The NRC administrative staff members are presented with awards during a reception in their honor.

Bottom: An NRC staff member makes comments during an office all-hands meeting.

ABBREVIATIONS USED

ABWR	Advanced Boiling-Water Reactor
AC	Allis Chalmers
ACRS	Advisory Committee of Radioactive Safety
AE	architect-engineer
AEC	Atomic Energy Commission (U.S.)
AEP	American Nuclear Power Company's Buchanan engineering offices
AGN	Solid homogeneous core (Aerojet-General Nucleonics)
AI	Atomics International
B&R	Burns & Roe
B&W	Babcock & Wilcox
BECH	Bechtel
BALD	Baldwin Associates
BLH	Baldwin Lima Hamilton
BRRT	Brown & Root
BWR	boiling-water reactor
CE	Combustion Engineering
CO	Company
CoC	Certificate of Compliance
COMM. OP.	date of commercial operation
CON TYPE	containment type
DRYAMB	dry, ambient pressure
DRYSUB	dry, subatmospheric
ICECND	wet, ice condenser
MARK 1	*wet, Mark I*
MARK 2	*wet, Mark II*
MARK 3	*wet, Mark III*
CP	construction permit
CP ISSUED	date of construction permit issuance
CVP	Civil Penalties
CVTR	Carolinas-Virginia Tube Reactor
CWE	Commonwealth Edison Company
CY	Calendar year
DANI	Daniel International
DBDB	Duke & Bechtel
DC	Design Certification
DOE	Department of Energy (U.S.)
DOT	Department of Transportation
DUKE	Duke Power Company
EBSO	Ebasco
EIA	Energy Information Administration (DOE)
EIS	Environment Impact Statement
EPR	Evolutionary Power Reactor
EPZ	Emergency Planning Zone
ERO	Emergency Response Organization

EVESR	ESADA (Empire States Atomic Development Associates) Vallecitos Experimental Superheat Reactor
EXP. DATE	expiration date of operating license
FBR	fast breeder reactor
FLUR	Fluor Pioneer
FR	*Federal Register*
FW	Foster Wheeler
FY	fiscal year
G&H	Gibbs & Hill
GA	General Atomic
GCR	gas-cooled reactor
GEH	General Electric-Hitachi Nuclear Energy
GEIS	Generic Environment Impact Statement
GETR	General Electric Test Reactor
GHDR	Gibbs & Hill & Durham & Richardson
GIL	Gilbert Associates
GL	General License
GPC	Georgia Power Company
GWe	gigawatt(s) electrical
HTG	high-temperature gas (reactor)
HWR	pressurized heavy-water reactor
INES	International Nuclear Event Scale
IRRS	IAEA Integrated Regulatory Review Service
ISFSI	Independent Spent Fuel Storage Installation
JONES	J.A. Jones
KAIS	Kaiser Engineers
KI	Potassium Iodine
kW	kilowatt(s)
LES	Louisiana Energy Services
LLP	B&W lowered loop
LMFB	liquid metal fast breeder (reactor)
LR ISSUED	license renewal issued
LWGR	graphite-moderated light-water reactor
MW	megawatt(s)
MWe	megawatt(s) electrical
MWh	megawatthour(s)
MWt	megawatt(s) thermal
NIAG	Niagara Mohawk Power Corporation
NISA	Japanese Nuclear and Industrial Safety Agency
NOV	notices of violation
NOVF	notices of violation associated with inspection findings

NOVSL	notices of violation for severity level	S&L	Sargent & Lundy
NRC	Nuclear Regulatory Commission (U.S.)	S&W	Stone & Webster
		SCF	sodium-cooled fast (reactor)
NSP	Northern States Power Company	SCGM	sodium-cooled, graphite-moderated (reactor)
NSSS	nuclear steam system supplier and design type	SDP	significance determination process
GE 2	*GE Type 2*	SGEC	architect for Vogtle
GE 3	*GE Type 3*	SI	systéme internationale (d'unités) (International System of Units)
GE 4	*GE Type 4*		
GE 5	*GE Type 5*	SL	Site Licenses
GE 6	*GE Type 6*	SOARCA	State-of-the-Art Consequence Analysis
WEST 2LP	*Westinghouse Two-Loop*		
WEST 3LP	*Westinghouse Three-Loop*	SSI	Southern Services Incorporated
WEST 4LP	*Westinghouse Four-Loop*	STARS	Strategic Teaming and Resource Sharing Group
OCM	organically cooled and moderated		
OL	operating license	STP	South Texas Project
OL ISSUED	date of latest full power operating license	TMI-2	Three Mile Island Unit 2
		TRACE	Reactor systems codes
PG&E	Pacific Gas & Electric Company	TRIGA	Training Reactor and Isotopes Production, General Atomics
PHWR	pressurized heavy-water-moderated and cooled (reactor)		
		TVA	Tennessee Valley Authority
PRA	Probabilistic risk assessment	UE&C	United Engineers & Constructors
PSE	Pioneer Services and Engineering	USEC	U.S. Enrichment Corporation
PSEG	Public Service Electric and Gas Company	US-APWR	United State Version of Advanced Pressurized Water Reactor
PTHW	pressure tube heavy water		
PUBS	Public Service Electric and Gas Company	VBWR	Vallecitos Boiling-Water Reactor
PWR	pressurized-water reactor	WDCO	Westinghouse Development Corporation
RTR	Research and Test Reactors	WEST	Westinghouse Electric

State and Territory Abbreviations

Alabama	AL	Kentucky	KY	Ohio	OH
Alaska	AK	Louisiana	LA	Oklahoma	OK
Arizona	AZ	Maine	ME	Oregon	OR
Arkansas	AR	Maryland	MD	Pennsylvania	PA
California	CA	Massachusetts	MA	Puerto Rico	PR
Colorado	CO	Michigan	MI	Rhode Island	RI
Connecticut	CT	Minnesota	MN	South Carolina	SC
Delaware	DE	Mississippi	MS	South Dakota	SD
District of Columbia	DC	Missouri	MO	Tennessee	TN
Florida	FL	Montana	MT	Texas	TX
Georgia	GA	Nebraska	NE	Utah	UT
Guam	GU	Nevada	NV	Vermont	VT
Hawaii	HI	New Hampshire	NH	Virgin Islands	VI
Idaho	ID	New Jersey	NJ	Virginia	VA
Illinois	IL	New Mexico	NM	Washington	WA
Indiana	IN	New York	NY	West Virginia	WV
Iowa	IA	North Carolina	NC	Wisconsin	WI
Kansas	KS	North Dakota	ND	Wyoming	WY

ABBREVIATIONS USED IN APPENDICES

APPENDIX A
U.S. Commercial Nuclear Power Reactors
Operating Reactors

Plant Name, Unit Number / Licensee / Location / Docket Number / NRC Web Page Address	NRC Region	Con Type / NSSS / Architect Engineer / Constructor	Licensed MWt	CP Issued / OL Issued / Comm. Op. / LR Issued / Exp. Date	2005–2010** Capacity Factor (Percent)
Arkansas Nuclear One, Unit 1 Entergy Operations, Inc. London, AR (6 miles NW of Russellville, AR) 050-00313 www.nrc.gov/info-finder/reactor/ano1.html	IV	PWR-DRYAMB B&W LLP BECH BECH	2,568	12/06/1968 05/21/1974 12/19/1974 06/20/2001 05/20/2034	78 102 94 83 99 90
Arkansas Nuclear One, Unit 2 Entergy Operations, Inc. London, AR (6 miles NW of Russellville, AR) 050-00368 www.nrc.gov/info-finder/reactor/ano2.html	IV	PWR-DRYAMB CE BECH BECH	3,026	12/06/1972 09/01/1978 03/26/1980 06/30/2005 07/17/2038	91 91 99 91 90 97
Beaver Valley Power Station, Unit 1 FirstEnergy Nuclear Operating Co. Shippingport, PA (17 miles W of McCandless, PA) 050-00334 www.nrc.gov/info-finder/reactor/bv1.html	I	PWR-DRYAMB WEST 3LP S&W S&W	2,900	06/26/1970 07/02/1976 10/01/1976 11/05/2009 01/29/2036	101 78 95 101 92 91
Beaver Valley Power Station, Unit 2 FirstEnergy Nuclear Operating Co. Shippingport, PA (17 miles W of McCandless, PA) 050-00412 www.nrc.gov/info-finder/reactor/bv2.html	I	PWR-DRYAMB WEST 3LP S&W S&W	2,900	05/03/1974 08/14/1987 11/17/1987 11/05/2009 05/27/2047	93 87 103 87 84 102
Braidwood Station, Unit 1 Exelon Generation Co., LLC Braceville, IL (20 miles SW of Joilet, IL) 050-00456 www.nrc.gov/info-finder/reactor/brai1.html	III	PWR-DRYAMB WEST 4LP S&L CWE	3,586.6	12/31/1975 07/02/1987 07/29/1988 N/A 10/17/2026	100 96 92 101 95 89
Braidwood Station, Unit 2 Exelon Generation Co., LLC Braceville, IL (20 miles SW of Joilet, IL) 050-00457 www.nrc.gov/info-finder/reactor/brai2.html	III	PWR-DRYAMB WEST 4LP S&L CWE	3,586.6	12/31/1975 05/20/1988 10/17/1988 N/A 12/18/2027	94 95 100 92 93 99
Browns Ferry Nuclear Plant, Unit 1 Tennessee Valley Authority Limestone County, AL (10 miles S of Athens, AL) 050-00259 www.nrc.gov/info-finder/reactor/bf1.html	II	BWR-MARK 1 GE 4 TVA TVA	3,458	05/10/1967 12/20/1973 08/01/1974 05/04/2006 12/20/2033	– – 49 88 94 86

U.S. Commercial Nuclear Power Reactors
Operating Reactors (continued)

Plant Name, Unit Number Licensee Location Docket Number NRC Web Page Address	NRC Region	Con Type NSSS Architect Engineer Constructor	Licensed MWt	CP Issued OL Issued Comm. Op. LR Issued Exp. Date	2005– 2010** Capacity Factor (Percent)
Browns Ferry Nuclear Plant, Unit 2 Tennessee Valley Authority Limestone County, AL (10 miles S of Athens, AL) 050-00260 www.nrc.gov/info-finder/reactor/bf2.html	II	BWR-MARK 1 GE 4 TVA TVA	3,458	05/10/1967 06/28/1974 03/01/1975 05/04/2006 06/28/2034	90 94 78 98 94 91
Browns Ferry Nuclear Plant, Unit 3 Tennessee Valley Authority Limestone County, AL (10 miles S of Athens, AL) 050-00296 www.nrc.gov/info-finder/reactor/bf3.html	II	BWR-MARK 1 GE 4 TVA TVA	3,458	07/31/1968 07/02/1976 03/01/1977 05/04/2006 07/02/2036	94 89 93 81 95 81
Brunswick Steam Electric Plant, Unit 1 Carolina Power & Light Co. Southport, NC (30 miles S of Wilmington, NC) 050-00325 www.nrc.gov/info-finder/reactor/bru1.html	II	BWR-MARK 1 GE 4 UE&C BRRT	2,923	02/07/1970 09/08/1976 03/18/1977 06/26/2006 09/08/2036	94 87 96 85 98 83
Brunswick Steam Electric Plant, Unit 2 Carolina Power & Light Co. Southport, NC (30 miles S of Wilmington, NC) 050-00324 www.nrc.gov/info-finder/reactor/bru2.html	II	BWR-MARK 1 GE 4 UE&C BRRT	2,923	02/07/1970 12/27/1974 11/03/1975 06/26/2006 12/27/2034	86 90 87 95 80 99
Byron Station, Unit 1 Exelon Generation Co., LLC Byron, IL (17 miles SW of Rockford, IL) 050-00454 www.nrc.gov/info-finder/reactor/byro1.html	III	PWR-DRYAMB WEST 4LP S&L CWE	3,586.6	12/31/1975 02/14/1985 09/16/1985 N/A 10/31/2024	94 91 98 95 94 101
Byron Station, Unit 2 Exelon Generation Co., LLC Byron, IL (17 miles SW of Rockford, IL) 050-00455 www.nrc.gov/info-finder/reactor/byro2.html	III	PWR-DRYAMB WEST 4LP S&L CWE	3,586.6	12/31/1975 01/30/1987 08/02/1987 N/A 11/06/2026	96 102 89 96 102 96
Callaway Plant Union Electric Co. Fulton, MO (25 miles NE of Jefferson City, MO) 050-00483 www.nrc.gov/info-finder/reactor/call.html	IV	PWR-DRYAMB WEST 4LP BECH DANI	3,565	04/16/1976 10/18/1984 12/19/1984 N/A 10/18/2024	77 97 90 90 98 86

U.S. Commercial Nuclear Power Reactors
Operating Reactors (continued)

Plant Name, Unit Number Licensee Location Docket Number NRC Web Page Address	NRC Region	Con Type NSSS Architect Engineer Constructor	Licensed MWt	CP Issued OL Issued Comm. Op. LR Issued Exp. Date	2005– 2010** Capacity Factor (Percent)
Calvert Cliffs Nuclear Power Plant, Unit 1 Calvert Cliffs Nuclear Power Plant, LLC Lusby, MD (40 miles S of Annapolis, MD) 050-00317 www.nrc.gov/info-finder/reactor/calv1.html	I	PWR-DRYAMB CE BECH BECH	2,737	07/07/1969 07/31/1974 05/08/1975 03/23/2000 07/31/2034	100 84 99 93 98 90
Calvert Cliffs Nuclear Power Plant, Unit 2 Calvert Cliffs Nuclear Power Plant, LLC Lusby, MD (40 miles S of Annapolis, MD) 050-00318 www.nrc.gov/info-finder/reactor/calv2.html	I	PWR-DRYAMB CE BECH BECH	2,737	07/07/1969 08/13/1976 04/01/1977 03/23/2000 08/13/2036	94 98 90 99 93 97
Catawba Nuclear Station, Unit 1 Duke Energy Carolinas, LLC York, SC (18 miles S of Charlotte, NC) 050-00413 www.nrc.gov/info-finder/reactor/cat1.html	II	PWR-ICECND WEST 4LP DUKE DUKE	3,411	08/07/1975 01/17/1985 06/29/1985 12/05/2003 12/05/2043	93 82 102 89 91 100
Catawba Nuclear Station, Unit 2 Duke Energy Carolinas, LLC York, SC (18 miles S of Charlotte, NC) 050-00414 www.nrc.gov/info-finder/reactor/cat2.html	II	PWR-ICECND WEST 4LP DUKE DUKE	3,411	08/07/1975 05/15/1986 08/19/1986 12/05/2003 12/05/2043	102 89 84 103 90 92
Clinton Power Station, Unit 1 Exelon Generation Co., LLC Clinton, IL (23 miles SSE of Bloomington, IL) 050-00461 www.nrc.gov/info-finder/reactor/clin.html	III	BWR-MARK 3 GE 6 S&L BALD	3,473	02/24/1976 04/17/1987 11/24/1987 N/A 09/29/2026	94 90 101 99 97 92
Columbia Generating Station Energy Northwest Benton County, WA (12 miles NW of Richland, WA) 050-00397 www.nrc.gov/info-finder/reactor/wash2.html	IV	BWR-MARK 2 GE 5 B&R BECH	3,486	03/19/1973 04/13/1984 12/13/1984 N/A 12/20/2023	83 94 82 93 67 95
Comanche Peak Nuclear Power Plant, Unit 1 Luminant Generation Co., LLC Glen Rose, TX (40 miles SW of Fort Worth, TX) 050-00445 www.nrc.gov/info-finder/reactor/cp1.html	IV	PWR-DRYAMB WEST 4LP G&H BRRT	3,612	12/19/1974 04/17/1990 08/13/1990 N/A 02/08/2030	92 102 185 96 100 91

U.S. Commercial Nuclear Power Reactors
Operating Reactors (continued)

Plant Name, Unit Number Licensee Location Docket Number NRC Web Page Address	NRC Region	Con Type NSSS Architect Engineer Constructor	Licensed MWt	CP Issued OL Issued Comm. Op. LR Issued Exp. Date	2005– 2010** Capacity Factor (Percent)
Comanche Peak Nuclear Power Plant, Unit 2 Luminant Generation Company, LLC Glen Rose, TX (40 miles SW of Fort Worth, TX) 050-00446 www.nrc.gov/info-finder/reactor/cp2.html	IV	PWR-DRYAMB WEST 4LP BECH BRRT	3,612	12/19/1974 04/06/1993 08/03/1993 N/A 02/02/2033	92 95 102 95 94 104
Cooper Nuclear Station Nebraska Public Power District Brownville, NE (23 miles S of Nebraska City, NE) 050-00298 www.nrc.gov/info-finder/reactor/cns.html	IV	BWR-MARK 1 GE 4 B&R B&R	2,419	06/04/1968 01/18/1974 07/01/1974 11/29/2010 01/18/2034	89 89 100 90 72 100
Crystal River Nuclear Generating Plant, Unit 3 Florida Power Corp. Crystal River, FL (80 miles N of Tampa, FL) 050-00302 www.nrc.gov/info-finder/reactor/cr3.html	II	PWR-DRYAMB B&W LLP GIL JONES	2,609	09/25/1968 12/03/1976 03/13/1977 N/A 12/03/2016	87 85 91 95 95 0
Davis-Besse Nuclear Power Station, Unit 1 FirstEnergy Nuclear Operating Co. Oak Harbor, OH (21 miles ESE of Toledo, OH) 050-00346 www.nrc.gov/info-finder/reactor/davi.html	III	PWR-DRYAMB B&W LLP BECH B&W	2,817	03/24/1971 04/22/1977 07/31/1978 N/A 04/22/2017	94 82 99 97 99 66
Diablo Canyon Nuclear Power Plant, Unit 1 Pacific Gas & Electric Co. Avila Beach, CA (12 miles SW of San Luis Obispo, CA) 050-00275 www.nrc.gov/info-finder/reactor/diab1.html	IV	PWR-DRYAMB WEST 4LP PG&E PG&E	3,411	4/23/1968 11/02/1984 05/07/1985 N/A 11/02/2024	87 101 90 98 84 88
Diablo Canyon Nuclear Power Plant, Unit 2 Pacific Gas & Electric Co. Avila Beach, CA 12 miles SW of San Luis Obispo, CA) 050-00323 www.nrc.gov/info-finder/reactor/diab2.html	IV	PWR-DRYAMB WEST 4LP PG&E PG&E	3,411	12/09/1970 08/26/1985 03/13/1986 N/A 08/26/2025	99 87 99 74 84 100
Donald C. Cook Nuclear Plant, Unit 1 Indiana Michigan Power Co. Bridgman, MI (13 miles S of Benton Harbor, MI) 050-00315 www.nrc.gov/info-finder/reactor/cook1.html	III	PWR-ICECND WEST 4LP AEP AEP	3,304	03/25/1969 10/25/1974 08/28/1975 08/30/2005 10/25/2034	91 81 103 64 3 88

APPENDIX A

APPENDIX A
U.S. Commercial Nuclear Power Reactors
Operating Reactors (continued)

Plant Name, Unit Number / Licensee / Location / Docket Number / NRC Web Page Address	NRC Region	Con Type / NSSS / Architect Engineer / Constructor	Licensed MWt	CP Issued / OL Issued / Comm. Op. / LR Issued / Exp. Date	2005–2010** Capacity Factor (Percent)
Donald C. Cook Nuclear Plant, Unit 2 Indiana Michigan Power Co. Bridgman, MI (13 miles S of Benton Harbor, MI) 050-00316 www.nrc.gov/info-finder/reactor/cook2.html	III	PWR-ICECND WEST 4LP AEP AEP	3,468	03/25/1969 12/23/1977 07/01/1978 08/30/2005 12/23/2037	100 89 86 101 87 84
Dresden Nuclear Power Station, Unit 2 Exelon Generation Co., LLC Morris, IL (25 miles SW of Joliet, IL) 050-00237 www.nrc.gov/info-finder/reactor/dres2.html	III	BWR-MARK 1 GE 3 S&L UE&C	2,957	01/10/1966 02/20/1991A 06/09/1970 10/28/2004 12/22/2029	87 96 92 98 91 102
Dresden Nuclear Power Station, Unit 3 Exelon Generation Co., LLC Morris, IL (25 miles SW of Joliet, IL) 050-00249 www.nrc.gov/info-finder/reactor/dres3.html	III	BWR-MARK 1 GE 3 S&L UE&C	2,957	10/14/1966 01/12/1971 11/16/1971 10/28/2004 01/12/2031	93 94 100 93 97 90
Duane Arnold Energy Center NextEra Energy Duane Arnold, LLC Palo, IA (8 miles NW of Cedar Rapids, IA) 050-00331 www.nrc.gov/info-finder/reactor/duan.html	III	BWR-MARK 1 GE 4 BECH BECH	1,912	06/22/1970 02/22/1974 02/01/1975 12/16/2010 02/21/2034	89 100 89 103 92 89
Edwin I. Hatch Nuclear Plant, Unit 1 Southern Nuclear Operating Co. Baxley, GA (20 miles S of Vidalia, GA) 050-00321 www.nrc.gov/info-finder/reactor/hat1.html	II	BWR-MARK 1 GE 4 BECH GPC	2,804	09/30/1969 10/13/1974 12/31/1975 01/15/2002 08/06/2034	91 84 98 84 94 85
Edwin I. Hatch Nuclear Plant, Unit 2 Southern Nuclear Operating Co., Inc. Baxley, GA (20 miles S of Vidalia, GA) 050-00366 www.nrc.gov/info-finder/reactor/hat2.html	II	BWR-MARK 1 GE 4 BECH GPC	2,804	12/27/1972 06/13/1978 09/05/1979 01/15/2002 06/13/2038	87 99 87 96 67 96
Fermi, Unit 2 The Detroit Edison Co. Newport, MI (25 miles NE of Toledo, OH) 050-00341 www.nrc.gov/info-finder/reactor/ferm2.html	III	BWR-MARK 1 GE 4 S&L DANI	3,430	09/26/1972 07/15/1985 01/23/1988 N/A 03/20/2025	90 76 85 98 75 80

A: AEC issued a provisional OL on 12/22/1969, allowing commercial operation. The NRC issued a full-term OL on 03/20/1991.

U.S. Commercial Nuclear Power Reactors
Operating Reactors (continued)

Plant Name, Unit Number Licensee Location Docket Number NRC Web Page Address	NRC Region	Con Type NSSS Architect Engineer Constructor	Licensed MWt	CP Issued OL Issued Comm. Op. LR Issued Exp. Date	2005– 2010** Capacity Factor (Percent)
Fort Calhoun Station, Unit 1 Omaha Public Power District Ft. Calhoun, NE (19 miles N of Omaha, NE) 050-00285 www.nrc.gov/info-finder/reactor/fcs.html	IV	PWR-DRYAMB CE GHDR GHDR	1,500	06/07/1968 08/09/1973 09/26/1973 11/04/2003 08/09/2033	70 74 104 83 100 102
Grand Gulf Nuclear Station, Unit 1 Entergy Operations, Inc. Port Gibson, MS (20 miles S of Vicksburg, MS) 050-00416 www.nrc.gov/info-finder/reactor/gg1.html	IV	BWR-MARK 3 GE 6 BECH BECH	3,898	09/04/1974 11/01/1984 07/01/1985 N/A 11/01/2024	91 94 84 86 100 88
H.B. Robinson Steam Electric Plant, Unit 2 Carolina Power & Light Co. Hartsville, SC (26 miles NW of Florence, SC) 050-00261 www.nrc.gov/info-finder/reactor/rob2.html	II	PWR-DRYAMB WEST 3LP EBSO EBSO	2,339	04/13/1967 07/31/1970 03/07/1971 04/19/2004 07/31/2030	93 104 92 87 104 57
Hope Creek Generating Station, Unit 1 PSEG Nuclear, LLC Hancocks Bridge, NJ (18 miles SE of Wilmington, DE) 050-00354 www.nrc.gov/info-finder/reactor/hope.html	I	BWR-MARK 1 GE 4 BECH BECH	3,840	11/04/1974 07/25/1986 12/20/1986 07/20/2011 04/11/2046	86 92 87 108 95 93
Indian Point Nuclear Generating, Unit 2 Entergy Nuclear Operations, Inc. Buchanan, NY (24 miles N of New York City, NY) 050-00247 www.nrc.gov/info-finder/reactor/ip2.html	I	PWR-DRYAMB WEST 4LP UE&C WDCO	3,216	10/14/1966 09/28/1973 08/01/1974 N/A 09/28/2013	99 89 99 91 98 82
Indian Point Nuclear Generating, Unit 3 Entergy Nuclear Operations, Inc. Buchanan, NY (24 miles N of New York City, NY) 050-00286 www.nrc.gov/info-finder/reactor/ip3.html	I	PWR-DRYAMB WEST 4LP UE&C WDCO	3,216	08/13/1969 12/12/1975 08/30/1976 N/A 12/12/2015	90 100 87 107 85 99
James A. FitzPatrick Nuclear Power Plant Entergy Nuclear Operations, Inc. Scriba, NY (6 miles NE of Oswego, NY) 050-00333 www.nrc.gov/info-finder/reactor/fitz.html	I	BWR-MARK 1 GE 4 S&W S&W	2,536	05/20/1970 10/17/1974 07/28/1975 09/08/2008 10/17/2034	95 91 93 89 99 85

U.S. Commercial Nuclear Power Reactors
Operating Reactors (continued)

Plant Name, Unit Number Licensee Location Docket Number NRC Web Page Address	NRC Region	Con Type NSSS Architect Engineer Constructor	Licensed MWt	CP Issued OL Issued Comm. Op. LR Issued Exp. Date	2005–2010** Capacity Factor (Percent)
Joseph M. Farley Nuclear Plant, Unit 1 Southern Nuclear Operating Co. Columbia, AL (18 miles S of Dothan, AL) 050-00348 www.nrc.gov/info-finder/reactor/far1.html	II	PWR-DRYAMB WEST 3LP SSI DANI	2,775	08/16/1972 06/25/1977 12/01/1977 05/12/2005 06/25/2037	99 86 88 97 90 88
Joseph M. Farley Nuclear Plant, Unit 2 Southern Nuclear Operating Co. Columbia, AL (18 miles S of Dothan, AL) 050-00364 www.nrc.gov/info-finder/reactor/far2.html	II	PWR-DRYAMB WEST 3LP SSI BECH	2,775	08/16/1972 03/31/1981 07/30/1981 05/12/2005 03/31/2041	84 101 87 90 96 88
Kewaunee Power Station Dominion Energy Kewaunee, Inc. Kewaunee, WI (27 miles SE of Green Bay, WI) 050-00305 www.nrc.gov/info-finder/reactor/kewa.html	III	PWR-DRYAMB WEST 2LP PSE PSE	1,772	08/06/1968 12/21/1973 06/16/1974 02/24/2011 12/21/2033	63 75 95 90 93 102
LaSalle County Station, Unit 1 Exelon Generation Co., LLC Marseilles, IL (11 miles SE of Ottawa, IL) 050-00373 www.nrc.gov/info-finder/reactor/lasa1.html	III	BWR-MARK 2 GE 5 S&L CWE	3,546	09/10/1973 04/17/1982 01/01/1984 N/A 04/17/2022	100 93 99 100 99 94
LaSalle County Station, Unit 2 Exelon Generation Co., LLC Marseilles, IL (11 miles SE of Ottawa, IL) 050-00374 www.nrc.gov/info-finder/reactor/lasa2.html	III	BWR-MARK 2 GE 5 S&L CWE	3,546	09/10/1973 12/16/1983 10/19/1984 N/A 12/16/2023	91 102 95 94 93 101
Limerick Generating Station, Unit 1 Exelon Generation Co., LLC Limerick, PA (21 miles NW of Philadelphia, PA) 050-00352 www.nrc.gov/info-finder/reactor/lim1.html	I	BWR-MARK 2 GE 4 BECH BECH	3,515	06/19/1974 08/08/1985 02/01/1986 N/A 10/26/2024	99 93 101 95 101 91
Limerick Generating Station, Unit 2 Exelon Generation Co., LLC Limerick, PA (21 miles NW of Philadelphia, PA) 050-00353 www.nrc.gov/info-finder/reactor/lim2.html	I	BWR-MARK 2 GE 4 BECH BECH	3,515	06/19/1974 08/25/1989 01/08/1990 N/A 06/22/2029	91 100 91 101 94 99

U.S. Commercial Nuclear Power Reactors
Operating Reactors (continued)

Plant Name, Unit Number Licensee Location Docket Number NRC Web Page Address	NRC Region	Con Type NSSS Architect Engineer Constructor	Licensed MWt	CP Issued OL Issued Comm. Op. LR Issued Exp. Date	2005– 2010** Capacity Factor (Percent)
McGuire Nuclear Station, Unit 1 Duke Energy Carolinas, LLC Huntersville, NC (17 miles N of Charlotte, NC) 050-00369 www.nrc.gov/info-finder/reactor/mcg1.html	II	PWR-ICECND WEST 4LP DUKE DUKE	3,411	02/23/1973 07/08/1981 12/01/1981 12/05/2003 06/12/2041	93 103 79 87 104 92
McGuire Nuclear Station, Unit 2 Duke Energy Carolinas, LLC Huntersville, NC (17 miles N of Charlotte, NC) 050-00370 www.nrc.gov/info-finder/reactor/mcg2.html	II	PWR-ICECND WEST 4LP DUKE DUKE	3,411	02/23/1973 05/27/1983 03/01/1984 12/05/2003 03/03/2043	89 87 103 90 94 104
Millstone Power Station, Unit 2 Dominion Nuclear Connecticut, Inc. Waterford, CT (3.2 miles SW of New London, CT) 050-00336 www.nrc.gov/info-finder/reactor/mill2.html	I	PWR-DRYAMB CE BECH BECH	2,700	12/11/1970 09/26/1975 12/26/1975 11/28/2005 07/31/2035	88 84 100 86 81 97
Millstone Power Station, Unit 3 Dominion Nuclear Connecticut, Inc. Waterford, CT (3.2 miles SW of New London, CT) 050-00423 www.nrc.gov/info-finder/reactor/mill3.html	I	PWR-DRYSUB WEST 4LP S&W S&W	3,650	08/09/1974 01/31/1986 04/23/1986 11/28/2005 11/25/2045	86 100 86 88 105 86
Monticello Nuclear Generating Plant, Unit 1 Northern States Power Company Monticello, MN (30 miles NW of Minneapolis, MN) 050-00263 www.nrc.gov/info-finder/reactor/mont.html	III	BWR-MARK GE 3 BECH BECH	1,775	06/19/1967 01/09/1981ᴮ 06/30/1971 11/08/2006 09/08/2030	89 101 84 97 83 94
Nine Mile Point Nuclear Station, Unit 1 Nine Mile Point Nuclear Station, LLC Scriba, NY (6 miles NE of Oswego, NY) 050-00220 www.nrc.gov/info-finder/reactor/nmp1.html	I	BWR-MARK 1 GE 2 NIAG S&W	1,850	04/12/1965 12/26/1974ᶜ 12/01/1969 10/31/2006 08/22/2029	85 98 88 98 92 97
Nine Mile Point Nuclear Station, Unit 2 Nine Mile Point Nuclear Station, LLC Scriba, NY (6 miles NE of Oswego, NY) 050-00410 www.nrc.gov/info-finder/reactor/nmp2.html	I	BWR-MARK 2 GE 5 S&W S&W	3,467	06/24/1974 07/02/1987 03/11/1988 10/31/2006 10/31/2046	100 90 92 90 99 89

B: AEC issued a provisional OL on 09/08/1970, allowing commercial operation. The NRC issued a full-term OL on 01/09/1981.

C: AEC issued a provisional OL on 08/22/1969, allowing commercial operation. The NRC issued a full-term OL on 12/26/1974.

APPENDIX A

U.S. Commercial Nuclear Power Reactors
Operating Reactors (continued)

Plant Name, Unit Number Licensee Location Docket Number NRC Web Page Address	NRC Region	Con Type NSSS Architect Engineer Constructor	Licensed MWt	CP Issued OL Issued Comm. Op. LR Issued Exp. Date	2005– 2010** Capacity Factor (Percent)
North Anna Power Station, Unit 1 Virginia Electric & Power Co. Louisa, VA (40 miles NW of Richmond, VA) 050-00338 www.nrc.gov/info-finder/reactor/na1.html	II	PWR-DRYSUB WEST 3LP S&W S&W	2,940	02/19/1971 04/01/1978 06/06/1978 03/20/2003 04/01/2038	95 88 89 101 92 86
North Anna Power Station, Unit 2 Virginia Electric & Power Co. Louisa, VA (40 miles NW of Richmond, VA) 050-00339 www.nrc.gov/info-finder/reactor/na2.html	II	PWR-DRYSUB WEST 3LP S&W S&W	2,940	02/19/1971 08/21/1980 12/14/1980 03/20/2003 08/21/2040	92 87 100 85 82 100
Oconee Nuclear Station, Unit 1 Duke Energy Carolinas, LLC Seneca, SC (30 miles W of Greenville, SC) 050-00269 www.nrc.gov/info-finder/reactor/oco1.html	II	PWR-DRYAMB B&W LLP DBDB DUKE	2,568	11/06/1967 02/06/1973 07/15/1973 05/23/2000 02/06/2033	91 79 99 84 85 100
Oconee Nuclear Station, Unit 2 Duke Energy Carolinas, LLC Seneca, SC (30 miles W of Greenville, SC) 050-00270 www.nrc.gov/info-finder/reactor/oco2.html	II	PWR-DRYAMB B&W LLP DBDB DUKE	2,568	11/06/1967 10/06/1973 09/09/1974 05/23/2000 10/06/2033	90 100 91 86 103 91
Oconee Nuclear Station, Unit 3 Duke Energy Carolinas, LLC Seneca, SC (30 miles W of Greenville, SC) 050-00287 www.nrc.gov/info-finder/reactor/oco3.html	II	PWR-DRYAMB B&W LLP DBDB DUKE	2,568	11/06/1967 07/19/1974 12/16/1974 05/23/2000 07/19/2034	98 91 87 102 94 91
Oyster Creek Nuclear Generating Station Exelon Generation Co., LLC Forked River, NJ (9 miles S of Toms River, NJ) 050-00219 www.nrc.gov/info-finder/reactor/oc.html	I	BWR-MARK 1 GE 2 B&R B&R	1,930	12/15/1964 07/02/1991D 12/23/1969 04/08/2009 04/09/2029	99 86 94 83 92 85
Palisades Nuclear Plant Entergy Nuclear Operations, Inc. Covert, MI (5 miles S of South Haven, MI) 050-00255 www.nrc.gov/info-finder/reactor/pali.html	III	PWR-DRYAMB CE BECH BECH	2,565.4	03/14/1967 03/24/1971 12/31/1971 01/17/2007 03/24/2031	79 98 86 99 90 92

D: AEC issued a provisional OL on 04/09/1969, allowing commercial operation. The NRC issued a full-term OL on 07/02/1991.

U.S. Commercial Nuclear Power Reactors
Operating Reactors (continued)

Plant Name, Unit Number Licensee Location Docket Number NRC Web Page Address	NRC Region	Con Type NSSS Architect Engineer Constructor	Licensed MWt	CP Issued OL Issued Comm. Op. LR Issued Exp. Date	2005– 2010** Capacity Factor (Percent)
Palo Verde Nuclear Generating Station, Unit 1 Arizona Public Service Company Wintersburg, AZ (50 miles W of Phoenix, AZ) 050-00528 www.nrc.gov/info-finder/reactor/palo1.html	IV	PWR-DRYAMB CE80-2L BECH BECH	3,990	05/25/1976 06/01/1985 01/28/1986 04/21/2011 06/01/2045	63 42 77 86 101 81
Palo Verde Nuclear Generating Station, Unit 2 Arizona Public Service Company Wintersburg, AZ (50 miles W of Phoenix, AZ) 050-00529 www.nrc.gov/info-finder/reactor/palo2.html	IV	PWR-DRYAMB CE80-2L BECH BECH	3,990	05/25/1976 04/24/1986 09/19/1986 04/21/2011 04/24/2046	82 85 95 74 83 101
Palo Verde Nuclear Generating Station, Unit 3 Arizona Public Service Company Wintersburg, AZ (50 miles W of Phoenix, AZ) 050-00530 www.nrc.gov/info-finder/reactor/palo3.html	IV	PWR-DRYAMB COMB CE80-2L BECH BECH	3,990	05/25/1976 11/25/1987 01/08/1988 04/21/2011 11/25/2047	84 86 64 97 83 89
Peach Bottom Atomic Power Station, Unit 2 Exelon Generation Co., LLC Delta, PA (17.9 miles S of Lancaster, PA) 050-00277 www.nrc.gov/info-finder/reactor/pb2.html	I	BWR-MARK 1 GE 4 BECH BECH	3,514	01/31/1968 10/25/1973 07/05/1974 05/07/2003 08/08/2033	98 93 101 89 101 92
Peach Bottom Atomic Power Station, Unit 3 Exelon Generation Co., LLC Delta, PA (17.9 miles S of Lancaster, PA) 050-00278 www.nrc.gov/info-finder/reactor/pb3.html	I	BWR-MARK 1 GE 4 BECH BECH	3,514	01/31/1968 07/02/1974 12/23/1974 05/07/2003 07/02/2034	91 102 93 93 89 100
Perry Nuclear Power Plant, Unit 1 FirstEnergy Nuclear Operating Co. Perry, OH (35 miles NE of Cleveland, OH) 050-00440 www.nrc.gov/info-finder/reactor/perr1.html	III	BWR-MARK 3 GE 6 GIL KAIS	3,758	05/03/1977 11/13/1986 11/18/1987 N/A 03/18/2026	71 97 75 98 67 98
Pilgrim Nuclear Power Station Entergy Nuclear Operations, Inc. Plymouth, MA (38 miles SE of Boston, MA) 050-00293 www.nrc.gov/info-finder/reactor/pilg.html	I	BWR-MARK 1 GE 3 BECH BECH	2,028	08/26/1968 06/08/1972 12/01/1972 N/A 06/08/2012	91 97 85 97 90 99

U.S. Commercial Nuclear Power Reactors
Operating Reactors (continued)

Plant Name, Unit Number Licensee Location Docket Number NRC Web Page Address	NRC Region	Con Type NSSS Architect Engineer Constructor	Licensed MWt	CP Issued OL Issued Comm. Op. LR Issued Exp. Date	2005– 2010** Capacity Factor (Percent)
Point Beach Nuclear Plant, Unit 1 NextEra Energy Point Beach, LLC Two Rivers, WI (13 miles NW of Manitowoc, WI) 050-00266 www.nrc.gov/info-finder/reactor/poin1.html	III	PWR-DRYAMB WEST 2LP BECH BECH	1,540	07/19/1967 10/05/1970 12/21/1970 12/22/2005 10/05/2030	81 100 85 87 98 88
Point Beach Nuclear Plant, Unit 2 NextEra Energy Point Beach, LLC Two Rivers, WI (13 miles NW of Manitowoc, WI) 050-00301 www.nrc.gov/info-finder/reactor/poin2.html	III	PWR-DRYAMB WEST 2LP BECH BECH	1,540	07/25/1968 03/08/1973^E 10/01/1972 12/22/2005 03/08/2033	72 91 99 89 84 96
Prairie Island Nuclear Generating Plant, Unit 1 Northern States Power Co.—Minnesota Welch, MN (28 miles SE of Minneapolis, MN) 050-00282 www.nrc.gov/info-finder/reactor/prai1.html	III	PWR-DRYAMB WEST 2LP FLUR NSP	1,677	06/25/1968 04/05/1974^F 12/16/1973 06/27/2011 08/09/2033	99 85 92 84 97 96
Prairie Island Nuclear Generating Plant, Unit 2 Northern States Power Co.—Minnesota Welch, MN (28 miles SE of Minneapolis, MN) 050-00306 www.nrc.gov/info-finder/reactor/prai2.html	III	PWR-DRYAMB WEST 2LP FLUR NSP	1,677	06/25/1968 10/29/1974 12/21/1974 06/27/2011 10/29/2034	84 84 93 85 97 86
Quad Cities Nuclear Power Station, Unit 1 Exelon Generation Co., LLC Cordova, IL (20 miles NE of Moline, IL) 050-00254 www.nrc.gov/info-finder/reactor/quad1.html	III	BWR-MARK 1 GE 3 S&L UE&C	2,957	02/15/1967 12/14/1972 02/18/1973 10/28/2004 12/14/2032	83 89 92 96 82 99
Quad Cities Nuclear Power Station, Unit 2 Exelon Generation Co., LLC Cordova, IL (20 miles NE of Moline, IL) 050-00265 www.nrc.gov/info-finder/reactor/quad2.html	III	BWR-MARK 1 GE 3 S&L UE&C	2,957	02/15/1967 12/14/1972 03/10/1973 10/28/2004 12/14/2032	93 86 99 86 91 92
River Bend Station, Unit 1 Entergy Operations, Inc. St. Francisville, LA (24 miles NW of Baton Rouge, LA) 050-00458 www.nrc.gov/info-finder/reactor/rbs1.html	IV	BWR-MARK 3 GE 6 S&W S&W	3,091	03/25/1977 11/20/1985 06/16/1986 N/A 08/29/2025	93 88 85 82 113 98

E: AEC issued a provisional OL on 11/18/1971. The NRC issued a full-term OL on 03/08/1973.
F: AEC issued a provisional OL on 08/09/1973. The NRC issued a full-term OL on 04/05/1974.

U.S. Commercial Nuclear Power Reactors
Operating Reactors (continued)

Plant Name, Unit Number Licensee Location Docket Number NRC Web Page Address	NRC Region	Con Type NSSS Architect Engineer Constructor	Licensed MWt	CP Issued OL Issued Comm. Op. LR Issued Exp. Date	2005– 2010** Capacity Factor (Percent)
R.E. Ginna Nuclear Power Plant R.E. Ginna Nuclear Power Plant, LLC Ontario, NY (20 miles NE of Rochester, NY) 050-00244 www.nrc.gov/info-finder/reactor/ginn.html	I	PWR-DRYAMB WEST 2LP GIL BECH	1,775	04/25/1966 09/19/1969 07/01/1970 05/19/2004 09/18/2029	92 95 113 109 91 97
St. Lucie Plant, Unit 1 Florida Power & Light Co. Jensen Beach, FL (10 miles SE of Ft. Pierce, FL) 050-00335 www.nrc.gov/info-finder/reactor/stl1.html	II	PWR-DRYAMB CE EBSO EBSO	2,700	07/01/1970 03/01/1976 12/21/1976 10/02/2003 03/01/2036	83 102 85 91 100 72
St. Lucie Plant, Unit 2 Florida Power & Light Co. Jensen Beach, FL (10 miles SE of Ft. Pierce, FL) 050-00389 www.nrc.gov/info-finder/reactor/stl2.html	II	PWR-DRYAMB CE EBSO EBSO	2,700	05/02/1977 06/10/1983 08/08/1983 10/02/2003 04/06/2043	86 82 70 99 80 100
Salem Nuclear Generating Station, Unit 1 PSEG Nuclear, LLC Hancocks Bridge, NJ (18 miles SE of Wilmington, DE) 050-00272 www.nrc.gov/info-finder/reactor/salm1.html	I	PWR-DRYAMB WEST 4LP PUBS UE&C	3,459	09/25/1968 12/01/1976 06/30/1977 06/30/2011 08/13/2036	92 99 89 91 99 85
Salem Nuclear Generating Station, Unit 2 PSEG Nuclear, LLC Hancocks Bridge, NJ (18 miles SE of Wilmington, DE) 050-00311 www.nrc.gov/info-finder/reactor/salm2.html	I	PWR-DRYAMB WEST 4LP PUBS UE&C	3,459	09/25/1968 05/20/1981 10/13/1981 06/30/2011 04/18/2040	90 92 98 83 93 98
San Onofre Nuclear Generating Station, Unit 2 Southern California Edison Co. San Clemente, CA (45 miles SE of Long Beach, CA) 050-00361 www.nrc.gov/info-finder/reactor/sano2.html	IV	PWR-DRYAMB CE BECH BECH	3,438	10/18/1973 02/16/1982 08/08/1983 N/A 02/16/2022	95 72 89 91 60 75
San Onofre Nuclear Generating Station, Unit 3 Southern California Edison Co. San Clemente, CA (45 miles SE of Long Beach, CA) 050-00362 www.nrc.gov/info-finder/reactor/sano3.html	IV	PWR-DRYAMB CE BECH BECH	3,438	10/18/1973 11/15/1982 04/01/1984 N/A 11/15/2022	100 72 94 69 104 72

APPENDIX A

U.S. Commercial Nuclear Power Reactors
Operating Reactors (continued)

Plant Name, Unit Number Licensee Location Docket Number NRC Web Page Address	NRC Region	Con Type NSSS Architect Engineer Constructor	Licensed MWt	CP Issued OL Issued Comm. Op. LR Issued Exp. Date	2005– 2010** Capacity Factor (Percent)
Seabrook Station, Unit 1 NextEra Energy Seabrook, LLC Seabrook, NH (13 miles S of Portsmouth, NH) 050-00443 www.nrc.gov/info-finder/reactor/seab1.html	I	PWR-DRYAMB WEST 4LP UE&C UE&C	3,648	07/07/1976 03/15/1990 08/19/1990 N/A 03/15/2030	89 69 99 89 81 100
Sequoyah Nuclear Plant, Unit 1 Tennessee Valley Authority Soddy-Daisy, TN (16 miles NE of Chattanooga, TN) 050-00327 www.nrc.gov/info-finder/reactor/seq1.html	II	PWR-ICECND WEST 4LP TVA TVA	3,455	05/27/1970 09/17/1980 07/01/1981 N/A 09/17/2020	100 90 87 101 89 84
Sequoyah Nuclear Plant, Unit 2 Tennessee Valley Authority Soddy-Daisy, TN (16 miles NE of Chattanooga, TN) 050-00328 www.nrc.gov/info-finder/reactor/seq2.html	II	PWR-ICECND WEST 4LP TVA TVA	3,455	05/27/1970 09/15/1981 06/01/1982 N/A 09/15/2021	90 90 100 89 89 97
Shearon Harris Nuclear Power Plant, Unit 1 Carolina Power & Light Co. New Hill, NC (20 miles SW of Raleigh, NC) 050-00400 www.nrc.gov/info-finder/reactor/har1.html	II	PWR-DRYAMB WEST 3LP EBSO DANI	2,900	01/27/1978 10/24/1986 05/02/1987 12/17/2008 10/24/2046	101 89 94 99 94 90
South Texas Project, Unit 1 STP Nuclear Operating Co. Bay City, TX (90 miles SW of Houston, TX) 050-00498 www.nrc.gov/info-finder/reactor/stp1.html	IV	PWR-DRYAMB WEST 4LP BECH EBSO	3,853	12/22/1975 03/22/1988 08/25/1988 N/A 08/20/2027	88 91 105 95 90 101
South Texas Project, Unit 2 STP Nuclear Operating Co. Bay City, TX (90 miles SW of Houston, TX) 050-00499 www.nrc.gov/info-finder/reactor/stp2.html	IV	PWR-DRYAMB WEST 4LP BECH EBSO	3,853	12/22/1975 03/28/1989 06/19/1989 N/A 12/15/2028	89 100 93 95 101 88
Surry Power Station, Unit 1 Virginia Electric and Power Co. Surry, VA (17 miles NW of Newport News, VA) 050-00280 www.nrc.gov/info-finder/reactor/sur1.html	II	PWR-DRYSUB WEST 3LP S&W S&W	2,857	06/25/1968 05/25/1972 12/22/1972 03/20/2003 05/25/2032	96 90 89 98 94 89
Surry Power Station, Unit 2 Virginia Electric and Power Co. Surry, VA (17 miles NW of Newport News, VA) 050-00281 www.nrc.gov/info-finder/reactor/sur2.html	II	PWR-DRYSUB WEST 3LP S&W S&W	2,857	06/25/1968 01/29/1973 05/01/1973 03/20/2003 01/29/2033	93 88 101 94 92 100

Plant Name, Unit Number Licensee Location Docket Number NRC Web Page Address	NRC Region	Con Type NSSS Architect Engineer Constructor	Licensed MWt	CP Issued OL Issued Comm. Op. LR Issued Exp. Date	2005– 2010** Capacity Factor (Percent)
Susquehanna Steam Electric Station, Unit 1 PPL Susquehanna, LLC Salem Township, Luzerne County, PA (70 miles NE of Harrisburg, PA) 050-00387 www.nrc.gov/info-finder/reactor/susq1.html	I	BWR-MARK 2 GE 4 BECH BECH	3,952	11/03/1973 07/17/1982 06/08/1983 11/24/2009 07/17/2042	95 86 95 89 101 80
Susquehanna Steam Electric Station, Unit 2 PPL Susquehanna, LLC Salem Township, Luzerne County, PA (70 miles NE of Harrisburg, PA) 050-00388 www.nrc.gov/info-finder/reactor/susq2.html	I	BWR-MARK 2 GE 4 BECH BECH	3,952	11/03/1973 03/23/1984 02/12/1985 11/24/2009 03/23/2044	89 93 88 100 90 96
Three Mile Island Nuclear Station, Unit 1 Exelon Generation Co., LLC Middletown, PA (10 miles SE of Harrisburg, PA) 050-00289 www.nrc.gov/info-finder/reactor/tmi1.html	I	PWR-DRYAMB B&W LLP GIL UE&C	2,568	05/18/1968 04/19/1974 09/02/1974 10/22/2009 04/19/2034	98 105 97 107 86 94
Turkey Point Nuclear Generating, Unit 3 Florida Power & Light Co. Homestead, FL (20 miles S of Miami, FL) 050-00250 www.nrc.gov/info-finder/reactor/tp3.html	II	PWR-DRYAMB WEST 3LP BECH BECH	2,300	04/27/1967 07/19/1972 12/14/1972 06/06/2002 07/19/2032	96 92 97 101 86 88
Turkey Point Nuclear Generating, Unit 4 Florida Power & Light Co. Homestead, FL (20 miles S of Miami, FL) 050-00251 www.nrc.gov/info-finder/reactor/tp4.html	II	PWR-DRYAMB WEST 3LP BECH BECH	2,300	04/27/1967 04/10/1973 09/07/1973 06/06/2002 04/10/2033	89 100 86 89 99 98
Vermont Yankee Nuclear Power Station Entergy Nuclear Operations, Inc. Vernon, VT (5 miles S of Brattleboro, VT) 050-00271 www.nrc.gov/info-finder/reactor/vy.html	I	BWR-MARK 1 GE 4 EBSO EBSO	1,912	12/11/1967 03/21/1972 11/30/1972 03/21/2011 03/21/2032	92 115 87 89 99 88
Virgil C. Summer Nuclear Station, Unit 1 South Carolina Electric & Gas Co. Jenkinsville, SC (26 miles NW of Columbia, SC) 050-00395 www.nrc.gov/info-finder/reactor/sum.html	II	PWR-DRYAMB WEST 3LP GIL DANI	2,900	03/21/1973 11/12/1982 01/01/1984 04/23/2004 08/06/2042	88 89 85 87 81 100
Vogtle Electric Generating Plant, Unit 1 Southern Nuclear Operating Co., Inc. Waynesboro, GA (26 miles SE of Augusta, GA) 050-00424 www.nrc.gov/info-finder/reactor/vog1.html	II	PWR-DRYAMB WEST 4LP SGEC GPC	3,625.6	06/28/1974 03/16/1987 06/01/1987 06/03/2009 01/16/2047	91 86 99 93 91 102

APPENDIX A

APPENDIX A
U.S. Commercial Nuclear Power Reactors
Operating Reactors (continued)

Plant Name, Unit Number Licensee Location Docket Number NRC Web Page Address	NRC Region	Con Type NSSS Architect Engineer Constructor	Licensed MWt	CP Issued OL Issued Comm. Op. LR Issued Exp. Date	2005– 2010** Capacity Factor (Percent)
Vogtle Electric Generating Plant, Unit 2 Southern Nuclear Operating Co., Inc. Waynesboro, GA (26 miles SE of Augusta, GA) 050-00425 www.nrc.gov/info-finder/reactor/vog2.html	II	PWR-DRYAMB WEST 4LP SBEC GPC	3,625.6	06/28/1974 03/31/1989 05/20/1989 06/03/2009 02/09/2049	85 92 83 88 101 93
Waterford Steam Electric Station, Unit 3 Entergy Operations, Inc. Killona, LA (25 miles W of New Orleans, LA) 050-00382 www.nrc.gov/info-finder/reactor/wat3.html	IV	PWR-DRYAMB COMB CE EBSO EBSO	3,716	11/14/1974 03/16/1985 09/24/1985 N/A 12/18/2024	78 92 98 89 87 100
Watts Bar Nuclear Plant, Unit 1 Tennessee Valley Authority Spring City, TN (60 miles SW of Knoxville, TN) 050-00390 www.nrc.gov/info-finder/reactor/wb1.html	II	PWR-ICECND WEST 4LP TVA TVA	3,459	01/23/1973 02/07/1996 05/27/1996 N/A 11/09/2035	90 68 102 82 94 99
Wolf Creek Generating Station, Unit 1 Wolf Creek Nuclear Operating Corp. Burlington, Coffey County, KS (28 miles SE of Emporia, KS) 050-00482 www.nrc.gov/info-finder/reactor/wc.html	IV	PWR-DRYAMB WEST 4LP BECH DANI	3,565	05/31/1977 06/04/1985 09/03/1985 11/20/2008 03/11/2045	99 86 92 102 83 86

Reactors Under Active Construction or Deferred Policy

Plant Name, Unit Number	NRC Region	Con Type	Licensed MWt	CP Issued	2005–2010**
Bellefonte Nuclear Power Station, Unit 1*** Tennessee Valley Authority (6 miles NE of Scottsboro, AL) 050-00438	II	PWR-DRYAMB B&W 205 TVA TVA	3,763	12/24/1974	N/A
Bellefonte Nuclear Power Station, Unit 2*** Tennessee Valley Authority (6 miles NE of Scottsboro, AL) 050-00439	II	PWR-DRYAMB B&W 205 TVA TVA	3,763	12/24/1974	N/A
Watts Bar Nuclear Plant, Unit 2**** Tennessee Valley Authority Spring City, TN (60 miles SW of Knoxville, TN) 050-00391	II	PWR-ICECND WEST 4LP TVA TVA	3,411	01/23/1973	

* Note: Plant names as identified on license as of May 31, 2011.

** Average capacity factor is listed in year order starting with 2005.

***Bellefonte Units 1 and 2 are under the Commission Policy Statement on Deferred Plants (52 FR 38077; October 14, 1987).

****Watts Bar Unit 2 is currently under active construction.

Source: NRC, with some data compiled from EIA/DOE

APPENDIX B
U.S. Commercial Nuclear Power Reactors
Permanently Shut Down—Formerly Licensed To Operate

Unit Location	Reactor Type MWt	NSSS Vendor	OL Issued Shut Down	Decommissioning Alternative Selected Current Status
Big Rock Point Charlevoix, MI	BWR 240	GE	05/01/1964 08/29/1997	DECON DECON Completed
GE Bonus* Punta Higuera, PR	BWR 50	CE	04/02/1964 06/01/1968	ENTOMB ENTOMB
CVTR** Parr, SC	PTHW 65	WEST	11/27/1962 01/01/1967	SAFSTOR SAFSTOR
Dresden 1 Morris, IL	BWR 700	GE	09/28/1959 10/31/1978	SAFSTOR SAFSTOR
Elk River* Elk River, MN	BWR 58	AC/S&L	11/06/1962 02/01/1968	DECON DECON Completed
Fermi 1 Newport, MI	SCF 200	CE	05/10/1963 09/22/1972	DECON DECON in Progress
Fort St. Vrain Platteville, CO	HTG 842	GA	12/21/1973 08/18/1989	DECON DECON Completed
GE VBWR Sunol, CA	BWR 50	GE	08/31/1957 12/09/1963	SAFSTOR SAFSTOR
Haddam Neck Meriden, CT	PWR 1,825	WEST	12/27/1974 12/05/1996	DECON DECON Completed
Hallam* Hallam, NE	SCGM 256	BLH	01/02/1962 09/01/1964	ENTOMB ENTOMB
NS Savannah Baltimore, MD	PWR 74	B&W	08/1965 11/1970	SAFSTOR SAFSTOR
Humboldt Bay 3 Eureka, CA	BWR 200	GE	08/28/1962 07/02/1976	DECON DECON In Progress
Indian Point 1 Buchanan, NY	PWR 615	B&W	03/26/1962 10/31/1974	SAFSTOR SAFSTOR
La Crosse Genoa, WI	BWR 165	AC	07/03/1967 04/30/1987	SAFSTOR SAFSTOR
Maine Yankee Wiscasset, ME	PWR 2,700	CE	06/29/1973 12/06/1996	DECON DECON Completed
Millstone 1 Waterford, CT	BWR 2,011	GE	10/31/1970 07/21/1998	SAFSTOR SAFSTOR
Pathfinder Sioux Falls, SD	BWR 190	AC	03/12/1964 09/16/1967	DECON DECON Completed
Peach Bottom 1 Delta, PA	HTG 115	GA	01/24/1966 10/31/1974	SAFSTOR SAFSTOR

APPENDIX B
U.S. Commercial Nuclear Power Reactors
Permanently Shut Down—Formerly Licensed To Operate (continued)

Unit Location	Reactor Type MWt	NSSS Vendor	OL Issued Shut Down	Decommissioning Alternative Selected Current Status
Piqua* Piqua, OH	OCM 46	AI	08/23/1962 01/01/1966	ENTOMB ENTOMB
Rancho Seco Herald, CA	PWR 2,772	B&W	08/16/1974 06/07/1989	DECON DECON Completed
San Onofre 1 San Clemente, CA	PWR 1,347	WEST	03/27/1967 11/30/1992	DECON DECON In Progress
Saxton Saxton, PA	PWR 23.5	WEST	11/15/1961 05/01/1972	DECON DECON Completed
Shippingport* Shippingport, PA	PWR 236	WEST	N/A 1982	DECON DECON Completed
Shoreham Wading River, NY	BWR 2,436	GE	04/21/1989 06/28/1989	DECON DECON Completed
Three Mile Island 2 Middletown, PA	PWR 2,770	B&W	02/08/1978 03/28/1979	(1)
Trojan Rainier, OR	PWR 3,411	WEST	11/21/1975 11/09/1992	DECON DECON Completed
Yankee-Rowe Rowe, MA	PWR 600	WEST	12/24/1963 10/01/1991	DECON DECON Completed
Zion 1 Zion, IL	PWR 3,250	WEST	10/19/1973 02/21/1997	DECON DECON In Progress
Zion 2 Zion, IL	PWR 3,250	WEST	11/14/1973 09/19/1996	DECON DECON In Progress

* AEC/DOE owned; not regulated by the U.S. Nuclear Regulatory Commission.

** Holds byproduct license from the State of South Carolina.

Notes: See Glossary for definitions of decommissioning alternatives (DECON, ENTOMB, SAFSTOR).

(1) Three Mile Island Unit 2 has been placed in a postdefueling monitored storage mode until Unit 1 permanently ceases operation, at which time both units are planned to be decommissioned.

Source: DOE Integrated Database for 1990; "U.S. Spent Fuel and Radioactive Waste, Inventories, Projections, and Characteristics" (DOE/RW-0006, Rev. 6), and U.S. Nuclear Regulatory Commission, "Nuclear Power Plants in the World," Edition 6

APPENDIX C
Canceled U.S. Commercial Nuclear Power Reactors

Unit Utility Location	Con Type MWe per Unit	Canceled Date Status
Allens Creek 1 Houston Lighting & Power Company 4 miles NW of Wallis, TX	BWR 1,150	1982 Under CP Review
Allens Creek 2 Houston Lighting & Power Company 4 miles NW of Wallis, TX	BWR 1,150	1976 Under CP Review
Atlantic 1 & 2 Public Service Electric & Gas Company Floating Plants off the Coast of NJ	PWR 1,150	1978 Under CP Review
Bailly 1 Northern Indiana Public Service Company 12 miles NNE of Gary, IN	BWR 645	1981 With CP
Barton 1 & 2 Alabama Power & Light 15 miles SE of Clanton, AL	BWR 1,159	1977 Under CP Review
Barton 3 & 4 Alabama Power & Light 15 miles SE of Clanton, AL	BWR 1,159	1975 Under CP Review
Black Fox 1 & 2 Public Service Company of Oklahoma 3.5 miles S of Inola, OK	BWR 1,150	1982 Under CP Review
Blue Hills 1 & 2 Gulf States Utilities Company SW tip of Toledo Bend Reservoir, TX	PWR 918	1978 Under CP Review
Callaway 2 Union Electric Company 25 miles ENE of Jefferson City, MO	PWR 1,150	1981 With CP
Cherokee 1 Duke Power Company 6 miles SSW of Blacksburg, SC	PWR 1,280	1983 With CP
Cherokee 2 & 3 Duke Power Company 6 miles SSW of Blacksburg, SC	PWR 1,280	1982 With CP
Clinch River Project Management Corp., DOE, TVA 23 miles W of Knoxville, in Oak Ridge, TN	LMFB 350	1983 Under CP Review

Canceled U.S. Commercial Nuclear Power Reactors (continued)

Unit Utility Location	Con Type MWe per Unit	Canceled Date Status
Clinton 2 Illinois Power Company 6 miles E of Clinton, IL	BWR 933	1983 With CP
Davis-Besse 2 & 3 Toledo Edison Company 21 miles ESE of Toledo, OH	PWR 906	1981 Under CP Review
Douglas Point 1 & 2 Potomac Electric Power Company Charles County, MD	BWR 1,146	1977 Under CP Review
Erie 1 & 2 Ohio Edison Company Berlin, OH	PWR 1,260	1980 Under CP Review
Forked River 1 Jersey Central Power & Light Company 2 miles S of Forked River, NJ	PWR 1,070	1980 With CP
Fort Calhoun 2 Omaha Public Power District 19 miles N of Omaha, NE	PWR 1,136	1977 Under CP Review
Fulton 1 & 2 Philadelphia Electric Company 17 miles S of Lancaster, PA	HTG 1,160	1975 Under CP Review
Grand Gulf 2 Entergy Nuclear Operations, Inc. 20 miles SW of Vicksburg, MS	BWR 1,250	1990 With CP
Greene County Power Authority of the State of NY 20 miles N of Kingston, NY	PWR 1,191	1980 Under CP Review
Greenwood 2 & 3 Detroit Edison Company Greenwood Township, MI	PWR 1,200	1980 Under CP Review
Hartsville A1 & A2 Tennessee Valley Authority 5 miles SE of Hartsville, TN	BWR 1,233	1984 With CP
Hartsville B1 & B2 Tennessee Valley Authority 5 miles SE of Hartsville, TN	BWR 1,233	1982 With CP

Canceled U.S. Commercial Nuclear Power Reactors (continued)

Unit Utility Location	Con Type MWe per Unit	Canceled Date Status
Haven 1 (formerly Koshkonong) Wisconsin Electric Power Company 4.2 miles SSW of Fort Atkinson, WI	PWR 900	1980 Under CP Review
Haven 2 (formerly Koshkonong) Wisconsin Electric Power Company 4.2 miles SSW of Fort Atkinson, WI	PWR 900	1978 Under CP Review
Hope Creek 2 Public Service Electric & Gas Company 18 miles SE of Wilmington, DE	BWR 1,067	1981 With CP
Jamesport 1 & 2 Long Island Lighting Company 65 miles E of New York City, NY	PWR 1,150	1980 With CP
Marble Hill 1 & 2 Public Service of Indiana 6 miles NE of New Washington, IN	PWR 1,130	1985 With CP
Midland 1 Consumers Power Company S of City of Midland, MI	PWR 492	1986 With CP
Midland 2 Consumers Power Company S of City of Midland, MI	PWR 818	1986 With CP
Montague 1 & 2 Northeast Nuclear Energy Company 1.2 miles SSE of Turners Falls, MA	BWR 1,150	1980 Under CP Review
New England 1 & 2 New England Power Company 8.5 miles E of Westerly, RI	PWR 1,194	1979 Under CP Review
New Haven 1 & 2 New York State Electric & Gas Corporation 3 miles NW of New Haven, NY	PWR 1,250	1980 Under CP Review
North Anna 3 Virginia Electric & Power Company 40 miles NW of Richmond, VA	PWR 907	1982 With CP
North Anna 4 Virginia Electric & Power Company 40 miles NW of Richmond, VA	PWR 907	1980 With CP

Canceled U.S. Commercial Nuclear Power Reactors (continued)

Unit Utility Location	Con Type MWe per Unit	Canceled Date Status
North Coast 1 Puerto Rico Water Resources Authority 4.7 miles ESE of Salinas, PR	PWR 583	1978 Under CP Review
Palo Verde 4 & 5 Arizona Public Service Company 36 miles W of Phoenix, AZ	PWR 1,270	1979 Under CP Review
Pebble Springs 1 & 2 Portland General Electric Company 55 miles WSW of Tri Cities (Kenewick-Pasco-Richland, WA), OR	PWR 1,260	1982 Under CP Review
Perkins 1, 2, & 3 Duke Power Company 10 miles N of Salisbury, NC	PWR 1,280	1982 Under CP Review
Perry 2 Cleveland Electric Illuminating Co. 35 miles NE of Cleveland, OH	BWR 1,205	1994 Under CP Review
Phipps Bend 1 & 2 Tennessee Valley Authority 15 miles SW of Kingsport, TN	BWR 1,220	1982 With CP
Pilgrim 2 Boston Edison Company 4 miles SE of Plymouth, MA	PWR 1,180	1981 Under CP Review
Pilgrim 3 Boston Edison Company 4 miles SE of Plymouth, MA	PWR 1,180	1974 Under CP Review
Quanicassee 1 & 2 Consumers Power Company 6 miles E of Essexville, MI	PWR 1,150	1974 Under CP Review
River Bend 2 Gulf States Utilities Company 24 miles NNW of Baton Rouge, LA	BWR 934	1984 With CP
Seabrook 2 Public Service Co. of New Hampshire 13 miles S of Portsmouth, NH	PWR 1,198	1988 With CP
Shearon Harris 2 Carolina Power & Light Company 20 miles SW of Raleigh, NC	PWR 900	1983 With CP

Canceled U.S. Commercial Nuclear Power Reactors (continued)

Unit Utility Location	Con Type MWe per Unit	Canceled Date Status
Shearon Harris 3 & 4 Carolina Power & Light Company 20 miles SW of Raleigh, NC	PWR 900	1981 With CP
Skagit/Hanford 1 & 2 Puget Sound Power & Light Company 23 miles SE of Bellingham, WA	PWR 1,277	1983 Under CP Review
Sterling Rochester Gas & Electric Corporation 50 miles E of Rochester, NY	PWR 1,150	1980 With CP
Summit 1 & 2 Delmarva Power & Light Company 15 miles SSW of Wilmington, DE	HTG 1,200	1975 Under CP Review
Sundesert 1 & 2 San Diego Gas & Electric Company 16 miles SW of Blythe, CA	PWR 974	1978 Under CP Review
Surry 3 & 4 Virginia Electric & Power Company 17 miles NW of Newport News, VA	PWR 882	1977 With CP
Tyrone 1 Northern States Power Company 8 miles NE of Durond, WI	PWR 1,150	1981 Under CP Review
Tyrone 2 Northern States Power Company 8 miles NE of Durond, WI	PWR 1,150	1974 With CP
Vogtle 3 & 4 Georgia Power Company 26 miles SE of Augusta, GA	PWR 1,113	1974 With CP
Washington Nuclear 1 Energy Northwest 10 miles E of Aberdeen, WA	PWR 1,266	1995 With CP
Washington Nuclear 3 Energy Northwest 16 miles E of Aberdeen, WA	PWR 1,242	1995 With CP
Washington Nuclear 4 Energy Northwest 10 miles E of Aberdeen, WA	PWR 1,218	1982 With CP

APPENDIX C
Canceled U.S. Commercial Nuclear Power Reactors (continued)

Unit Utility Location	Con Type MWe per Unit	Canceled Date Status
Washington Nuclear 5 Energy Northwest 16 miles E of Aberdeen, WA	PWR 1,242	1982 With CP
Yellow Creek 1 & 2 Tennessee Valley Authority 15 miles E of Corinth, MS	BWR 1,285	1984 With CP
Zimmer 1 Cincinnati Gas & Electric Company 25 miles SE of Cincinnati, OH	BWR 810	1984 With CP

Note: Cancellation is defined as public announcement of cancellation or written notification to the NRC. Only NRC-docketed applications are included. Status is the status of the application at the time of cancellation.

Source: DOE/EIA Commercial Nuclear Power 1991 (DOE/EIA-0438), Appendix E (page 105), and NRC

APPENDIX D
U.S. Commercial Nuclear Power Reactors by Parent Company

Utility	NRC-Abbreviated Reactor Unit Name
AmerenUE www.ameren.com	Callaway*
Arizona Public Service Company www.aps.com	Palo Verde 1, 2, & 3*
Constellation Energy www.constellation.com	Calvert Cliffs 1 & 2 Ginna Nine Mile Point 1 & 2
Detroit Edison Company www.dteenergy.com	Fermi 2
Dominion Generation www.dom.com	Kewaunee Millstone 2 & 3 North Anna 1 & 2 Surry 1 & 2
Duke Energy Carolinas, LLC www.duke-energy.com	Catawba 1 & 2 McGuire 1 & 2 Oconee 1, 2, & 3
Energy Northwest www.energy-northwest.com	Columbia
Entergy Nuclear Operations, Inc. www.entergy-nuclear.com	Arkansas Nuclear One 1 & 2 FitzPatrick Grand Gulf 1 Indian Point 2 & 3 Palisades Pilgrim 1 River Bend 1 Vermont Yankee Waterford 3
Exelon Corporation, LLC www.exeloncorp.com	Braidwood 1 & 2 Byron 1 & 2 Clinton Dresden 2 & 3 LaSalle 1 & 2 Limerick 1 & 2 Oyster Creek Peach Bottom 2 & 3 Quad Cities 1 & 2 Three Mile Island 1
FirstEnergy Nuclear Generating Corp. www.firstenergycorp.com	Beaver Valley 1 & 2 Davis-Besse Perry 1

U.S. Commercial Nuclear Power Reactors by Parent Company
(continued)

Utility	NRC-Abbreviated Reactor Unit Name
FPL Group, Inc. www.fplgroup.com	Duane Arnold Point Beach 1 & 2 Seabrook 1 St. Lucie 1 & 2 Turkey Point 3 & 4
Indiana Michigan Power Company www.indianamichiganpower.com	Cook 1 & 2
Luminant Generation Company, LLC www.luminant.com	Comanche Peak 1 & 2*
Nebraska Public Power District www.nppd.com	Cooper
Northern States Power, an Xcel Energy Operating Company www.xcelenergy.com	Monticello Prairie Island 1 & 2
Omaha Public Power District www.oppd.com	Fort Calhoun
Pacific Gas & Electric Company www.pge.com	Diablo Canyon 1 & 2*
PPL Susquehanna, LLC www.pplweb.com	Susquehanna 1 & 2
Progress Energy www.progress-energy.com	Brunswick 1 & 2 Crystal River 3 Robinson 2 Harris 1
PSEG Nuclear, LLC www.pseg.com	Hope Creek 1 Salem 1 & 2
South Carolina Electric & Gas Company www.sceg.com	Summer
Southern California Edison Company www.sce.com	San Onofre 2 & 3
Southern Nuclear Operating Company www.southerncompany.com	Hatch 1 & 2 Farley 1 & 2 Vogtle 1 & 2
STP Nuclear Operating Company www.stpnoc.com	South Texas Project 1 & 2*
Tennessee Valley Authority www.tva.gov	Browns Ferry 1, 2, & 3 Sequoyah 1 & 2 Watts Bar 1
Wolf Creek Nuclear Operating Corporation www.wcnoc.com	Wolf Creek 1*

*These plants have a joint program called the Strategic Teaming and Resource Sharing (STARS) group.
They share resources for refueling outages and develop some shared licensing applications.

Operating U.S. Nuclear Research and Test Reactors Regulated by the NRC

Licensee Location	Reactor Type OL Issued	Power Level (kW)	Licensee Number Docket Number
Aerotest San Ramon, CA	TRIGA (Indus) 07/02/1965	250	R-98 50-228
Armed Forces Radiobiology Research Institute Bethesda, MD	TRIGA 06/26/1962	1,100	R-84 50-170
Dow Chemical Company Midland, MI	TRIGA 07/03/1967	300	R-108 50-264
GE-Hitachi Sunol, CA	Tank 10/31/1957	100	R-33 50-73
Idaho State University Pocatello, ID	AGN-201 #103 10/11/1967	0.005	R-110 50-284
Kansas State University Manhattan, KS	TRIGA 10/16/1962	250	R-88 50-188
Massachusetts Institute of Technology Cambridge, MA	HWR Reflected 06/09/1958	6,000	R-37 50-20
National Institute of Standards & Technology Gaithersburg, MD	Nuclear Test 05/21/1970	20,000	TR-5 50-184
North Carolina State University Raleigh, NC	Pulstar 08/25/1972	1,000	R-120 50-297
Ohio State University Columbus, OH	Pool 02/24/1961	500	R-75 50-150
Oregon State University Corvallis, OR	TRIGA Mark II 03/07/1967	1,100	R-106 50-243
Pennsylvania State University State College, PA	TRIGA 07/08/1955	1,100	R-2 50-5
Purdue University West Lafayette, IN	Lockheed 08/16/1962	1	R-87 50-182
Reed College Portland, OR	TRIGA Mark I 07/02/1968	250	R-112 50-288
Rensselaer Polytechnic Institute Troy, NY	Critical Assembly 07/03/1964	0.1	CX-22 50-225
Rhode Island Atomic Energy Commission Narragansett, RI	GE Pool 07/23/1964	2,000	R-95 50-193

Operating U.S. Nuclear Research and Test Reactors
Regulated by the NRC (continued)

Licensee Location	Reactor Type OL Issued	Power Level (kW)	Licensee Number Docket Number
Texas A&M University College Station, TX	AGN-201M #106 08/26/1957	0.005	R-23 50-59
Texas A&M University College Station, TX	TRIGA 12/07/1961	1,000	R-128 50-128
U.S. Geological Survey Denver, CO	TRIGA Mark I 02/24/1969	1,000	R-113 50-274
University of California/Davis Sacramento, CA	TRIGA 08/13/1998	2,300	R-130 50-607
University of California/Irvine Irvine, CA	TRIGA Mark I 11/24/1969	250	R-116 50-326
University of Florida Gainesville, FL	Argonaut 05/21/1959	100	R-56 50-83
University of Maryland College Park, MD	TRIGA 10/14/1960	250	R-70 50-166
University of Massachusetts/Lowell Lowell, MA	GE Pool 12/24/1974	1,000	R-125 50-223
University of Missouri/Columbia Columbia, MO	Tank 10/11/1966	10,000	R-103 50-186
University of Missouri/Rolla Rolla, MO	Pool 11/21/1961	200	R-79 50-123
University of New Mexico Albuquerque, NM	AGN-201M #112 09/17/1966	0.005	R-102 50-252
University of Texas Austin, TX	TRIGA Mark II 01/17/1992	1,100	R-92 50-602
University of Utah Salt Lake City, UT	TRIGA Mark I 09/30/1975	100	R-126 50-407
University of Wisconsin Madison, WI	TRIGA 11/23/1960	1,000	R-74 50-156
Washington State University Pullman, WA	TRIGA 03/06/1961	1,000	R-76 50-27

APPENDIX F

U.S. Nuclear Research and Test Reactors
Under Decommissioning Regulated by the NRC

Licensee Location	Reactor Type Power Level (kW)	OL Issued Shutdown	Decommissioning Alternative Selected Current Status
General Atomics San Diego, CA	TRIGA Mark F 1,500	07/01/60 09/07/94	DECON SAFSTOR
General Atomics San Diego, CA	TRIGA Mark I 250	05/03/58 12/17/96	DECON SAFSTOR
General Electric Company Sunol, CA	GETR (Tank) 50,000	01/07/59 06/26/85	SAFSTOR SAFSTOR
General Electric Company Sunol, CA	EVESR 17,000	11/12/63 02/01/67	SAFSTOR SAFSTOR
National Aeronautics and Space Administration Sandusky, OH	Test 60,000	05/02/62 07/07/73	DECON DECON In Progress
National Aeronautics and Space Administration Sandusky, OH	Mockup 100	06/14/61 07/07/73	DECON DECON In Progress
University of Buffalo Buffalo, NY	Pulstar 2,000	03/24/61 07/23/96	DECON SAFSTOR
University of Illinois Urbana-Champaign, IL	TRIGA 1,500	07/22/69 04/12/99	SAFSTOR DECON In Progress
University of Michigan Ann Arbor, MI	Pool 2,000	09/13/57 01/29/04	DECON DECON In Progress
Veterans Administration Omaha, NE	TRIGA 20	06/26/59 11/05/01	DECON SAFSTOR
Worcester Polytechnic Institute Worcester, MA	GE 10	12/16/59 06/30/07	DECON DECON Pending
University of Arizona Tucson, AZ	TRIGA Mark I 110	12/05/58 05/18/10	DECON SAFSTOR

APPENDIX G
Industry Performance Indicators:
Annual Industry Averages, FYs 2001–2010

Indicator	2001	2002	2003	2004	2005	2006	2007	2008	2009	2010
Automatic Scrams	0.57	0.44	0.75	0.56	0.47	0.32	0.48	0.29	0.36	0.44
Safety System Actuations	0.19	0.18	0.41	0.24	0.38	0.22	0.25	0.14	0.23	0.18
Significant Events	0.07	0.05	0.07	0.04	0.05	0.03	0.02	0.02	0.00	0.09
Safety System Failures	0.82	0.88	0.96	0.78	0.99	0.59	0.68	0.69	0.67	0.89
Forced Outage Rate	3.00	1.70	3.04	1.88	2.44	1.47	1.43	1.34	2.21	1.74
Equipment-Forced Outage Rate	0.11	0.12	0.16	0.15	0.13	0.10	0.11	0.08	0.09	0.10
Collective Radiation Exposure	123	111	125	100	117	93	110	96	88	92
Drill/Exercise Performance	95	95	96	96	96	96	98	96	97	97
ERO Drill Participation	96	97	98	98	98	98	98	98	99	99
Alert and Notification System Reliability	99	99	99	99	99	99	99	100	100	100

APPENDIX H
Dry Spent Fuel Storage Designs:
NRC-Approved for Use by General Licensees

Vendor	Docket #	Storage Design Model
General Nuclear Systems, Inc.	72-1000	CASTOR V/21
NAC International, Inc.	72-1002	NAC S/T
	72-1003	NAC-C28 S/T
	72-1015	NAC-UMS
	72-1025	NAC-MPC
	72-1031	Magnastor
Holtec International	72-1008	HI-STAR 100
	72-1014	HI-STORM 100
BNG Fuel Solutions Corporation	72-1007	VSC-24
	72-1026	Fuel Solutions™ (WSNF-220, -221, -223)
		W-150 Storage Cask
		W-100 Transfer Cask
		W-21, W-74 Canisters
Transnuclear, Inc.	72-1005	TN-24
	72-1027	TN-68
	72-1021	TN-32, 32A, 32B
	72-1004	Standardized NUHOMS®-24P, -24PHB, -24PTH, -32PT, -32PTH1, -52B, -61BT, -61BTH
	72-1029	Standardized Advanced NUHOMS®-24PT1, -24PT4
	72-1030	NUHOMS® HD-32PTH

Data as of April 2011; see latest list on the NRC Web site at www.nrc.gov/waste/spent-fuel-storage/designs.html.

APPENDIX I
Dry Spent Fuel Storage Licensees

Name Licensee	License Type	Date Issued	Vendor	Storage Model	Docket #
Surry Virginia Electric & Power Company (Dominion Gen.)	SL	07/02/1986	General Nuclear Systems, Inc. Transnuclear, Inc. NAC International, Inc. Westinghouse, Inc.	CASTOR V/21 TN-32 NAC-128 CASTOR X/33 MC-10	72-2
	GL	08/06/2007	Transnuclear, Inc.	NUHOMS®-HD	72-55
H.B. Robinson Carolina Power & Light Company	SL GL	08/13/1986 09/06/2005	Transnuclear, Inc. Transnuclear, Inc.	NUHOMS®-7P NUHOMS®-24P	72-3 72-60
Oconee Duke Energy Company	SL GL	01/29/1990 03/05/1999	Transnuclear, Inc. Transnuclear, Inc.	NUHOMS®-24P NUHOMS®-24P	72-4 72-40
Fort St. Vrain* U.S. Department of Energy	SL	11/04/1991	FW Energy Applications, Inc.	Modular Vault Dry Store	72-9
Calvert Cliffs Calvert Cliffs Nuclear Power Plant, Inc.	SL	11/25/1992	Transnuclear, Inc.	NUHOMS®-24P NUHOMS®-32P	72-8
Palisades Entergy Nuclear Operations, Inc.	GL	05/11/1993	BNG Fuel Solutions Transnuclear, Inc.	VSC-24 NUHOMS®-32PT	72-7
Prairie Island Northern States Power Co., a Minnesota Corp.	SL	10/19/1993	Transnuclear, Inc.	TN-40	72-10
Point Beach FLP Energy Point Beach, LLC	GL	05/26/1996	BNG Fuel Solutions Transnuclear, Inc.	VSC-24 NUHOMS®-32PT	72-5
Davis-Besse FirstEnergy Nuclear Operating Company	GL	01/01/1996	Transnuclear, Inc.	NUHOMS®-24P	72-14
Arkansas Nuclear Entergy Nuclear Operations, Inc.	GL	12/17/1996	BNG Fuel Solutions Holtec International	VSC-24 HI-STORM 100	72-13
North Anna Virginia Electric & Power Company (Dominion Gen.)	SL GL	06/30/1998 03/10/2008	Transnuclear, Inc. Transnuclear, Inc.	TN-32 NUHOMS®-HD	72-16 72-56
Trojan Portland General Electric Corp.	SL	03/31/1999	Holtec International	HI-STORM 100	72-17

Dry Spent Fuel Storage Licensees (continued)

Name Licensee	License Type	Date Issued	Vendor	Storage Model	Docket #
Idaho National Lab TMI-2 Fuel Debris, U.S. Department of Energy	SL	03/19/1999	Transnuclear, Inc.	NUHOMS®-12T	72-20
Susquehanna PPL Susquehanna, LLC	GL	10/18/1999	Transnuclear, Inc.	NUHOMS®-52B NUHOMS®-61BT	72-28
Peach Bottom Exelon Generation Company, LLC	GL	06/12/2000	Transnuclear, Inc.	TN-68	72-29
Hatch Southern Nuclear Operating, Inc.	GL	07/06/2000	Holtec International	HI-STAR 100 HI-STORM 100	72-36
Dresden Exelon Generation Company, LLC	GL	07/10/2000	Holtec International	HI-STAR 100 HI-STORM 100	72-37
Rancho Seco Sacramento Municipal Utility District	SL	06/30/2000	Transnuclear, Inc.	NUHOMS®-24P	72-11
McGuire Duke Energy, LLC	GL	02/01/2001	Transnuclear, Inc.	TN-32	72-38
Big Rock Point Entergy Nuclear Operations, Inc.	GL	11/18/2002	BNG Fuel Solutions	Fuel Solutions™ W74	72-43
James A. FitzPatrick Entergy Nuclear Operations, Inc.	GL	04/25/2002	Holtec International	HI-STORM 100	72-12
Maine Yankee Maine Yankee Atomic Power Company	GL	08/24/2002	NAC International, Inc.	NAC-UMS	72-30
Columbia Generating Station Energy Northwest	GL	09/02/2002	Holtec International	HI-STORM 100	72-35
Oyster Creek AmerGen Energy Company, LLC.	GL	04/11/2002	Transnuclear, Inc.	NUHOMS®-61BT	72-15
Yankee Rowe Yankee Atomic Electric	GL	06/26/2002	NAC International, Inc.	NAC-MPC	72-31
Duane Arnold Next Era Energy Duane Arnold, LLC.	GL	09/01/2003	Transnuclear, Inc.	NUHOMS®-61BT	72-32

Dry Spent Fuel Storage Licensees (continued)

Name Licensee	License Type	Date Issued	Vendor	Storage Model	Docket #
Palo Verde Arizona Public Service Company	GL	03/15/2003	NAC International, Inc.	NAC-UMS	72-44
San Onofre Southern California Edison Company	GL	10/03/2003	Transnuclear, Inc.	NUHOMS®-24PT	72-41
Diablo Canyon Pacific Gas & Electric Co.	SL	03/22/2004	Holtec International	HI-STORM 100	72-26
Haddam Neck CT Yankee Atomic Power	GL	05/21/2004	NAC International, Inc.	NAC-MPC	72-39
Sequoyah Tennessee Valley Authority	GL	07/13/2004	Holtec International	HI-STORM 100	72-34
Idaho Spent Fuel Facility	SL	11/30/2004	Foster Wheeler Environmental Corp.	Concrete Vault	72-25
Humboldt Bay Pacific Gas & Electric Co.	SL	11/30/2005	Holtec International	HI-STORM 100HB	72-27
Private Fuel Storage Facility	SL	02/21/2006	Holtec International	HI-STORM 100	72-22
Browns Ferry Tennessee Valley Authority	GL	08/21/2005	Holtec International	HI-STORM 100S	72-52
Joseph M. Farley Southern Nuclear Operating Co.	GL	08/25/2005	Transnuclear, Inc.	NUHOMS®-32PT	72-42
Millstone Dominion Generation	GL	02/15/2005	Transnuclear, Inc.	NUHOMS®-32PT	72-47
Quad Cities Exelon Generation Company, LLC	GL	12/02/2005	Holtec International	HI-STORM 100S	72-53
River Bend Entergy Nuclear Operations, Inc.	GL	12/29/2005	Holtec International	HI-STORM 100S	72-49
Fort Calhoun Omaha Public Power District	GL	07/29/2006	Transnuclear, Inc.	NUHOMS®-32PT	72-54
Hope Creek/Salem PSEG, Nuclear, LLC	GL	11/10/2006	Holtec International	HI-STORM 100	72-48
Grand Gulf Entergy Nuclear Operations, Inc.	GL	11/18/2006	Holtec International	HI-STORM 100S	72-50

Dry Spent Fuel Storage Licensees (continued)

Name Licensee	License Type	Date Issued	Vendor	Storage Model	Docket #
Catawba Duke Energy Carolinas, LLC	GL	07/30/2007	NAC International, Inc.	NAC-UMS	72-45
Indian Point Entergy Nuclear Operations, Inc.	GL	01/11/2008	Holtec International	HI-STORM 100	72-51
St. Lucie Florida Power and Light Company	GL	03/14/2008	Transnuclear, Inc.	NUHOMS®-HD	72-61
Vermont Yankee Entergy Nuclear Operations, Inc.	GL	05/25/2008	Holtec International	HI-STORM100	72-59
Limerick Exelon Generation Co., LLC	GL	08/01/2008	Transnuclear, Inc.	NUHOMS®-61BT	72-65
Seabrook FPL Energy	GL	08/07/2008	Transnuclear, Inc.	NUHOMS®-HD-3PTM	72-61
Monticello Northern States Power Co.	GL	09/17/2008	Transnuclear, Inc.	NUHOMS®-61BT	72-58
Kewaunee Northern States Power Co.	GL	09/11/2009	Transnuclear, Inc.	NUHOMS®-39PT	72-64
Byron Exelon Generation Co., LLC	GL	09/09/2010	Holtec International	HI-STORM 100	72-68
La Salle Exelon Generation Co., LLC	GL	11/01/2010	Holtec International	HI-STORM100	72-70

*Fort St. Vrain is undergoing decommissioning and was transferred to DOE on June 4, 1999.

Note: NRC-abbreviated unit names.

APPENDIX J
Nuclear Power Units by Nation

Country	In Operation Number of Units	In Operation Capacity Net MWe	Under Construction or on Order Number of Units	Under Construction or on Order Capacity Net MWe	Nuclear Power Production GWe*	Shutdown
Argentina	2	935	1	692	6,692	0
Armenia	1	375	0	0	2,344	1ᴾ
Belgium	7	5,927	0	0	45,728	1ᴾ
Brazil	2	1,884	1	1,245	14,544	0
Bulgaria	2	1,906	2	1,906	15,249	4ᴾ
Canada	18	12,529	0	0	85,220	3ᴾ & 4ᴸ
China	14	11,058	27	27,230	76,817	0
Czech Republic	6	3,678	0	0	26,441	0
Finland	4	2,716	1	1,600	21,884	0
France	58	63,130	1	1,600	407,900	12ᴾ
Germany	17	20,490	0	0	133,012	19ᴾ
Hungary	4	1,889	0	0	14,803	0
India	20	4,391	5	3,564	20,481	0
Iran	0	0	1	915	0	0
Italy	0	0	0	0	0	4ᴾ
Japan	50	44,102	2	2,650	279,230	9ᴾ & 1ᴸ
Kazakhstan	0	0	0	0	0	1
Korea, South	21	18,698	5	5,560	141,894	0
Lithuania	0	0	0	0	0	2ᴾ
Mexico	2	1,300	0	0	5,596	0
Netherlands	1	482	0	0	3,755	1ᴾ
Pakistan	3	725	0	0	2,560	0
Romania	2	1,300	0	0	10,705	0
Russia	32	22,693	11	9,153	155,108	5ᴾ
Slovakia	4	1,816	2	782	13,534	3ᴾ
Slovenia	1	688	0	0	5,381	0
South Africa	2	1,800	0	0	12,100	0
Spain	8	7,514	0	0	59,256	2ᴾ
Sweden	10	9,298	0	0	55,100	3
Switzerland	5	3,263	0	0	25,200	1ᴾ

Nuclear Power Units by Nation (continued)

Country	In Operation Number of Units	In Operation Capacity Net MWe	Under Construction or on Order Number of Units	Under Construction or on Order Capacity Net MWe	Nuclear Power Production GWe*	Shutdown
Ukraine	15	13,107	2	1,900	83,800	4[P]
United Kingdom	19	10,137	0	0	56,440	26
United States	104	101,240	1	1,165	806,968	28
Total	440	374,093	64	65,562	2,587,739	129[P] & 5[L]

* Annual electrical power production for 2010

P = Permanent Shutdown
L = Long-term Shutdown

Note: Operable, under construction, or on order. Country's short-form name used. Rounded to the nearest whole number.

Source: IAEA Power Reactor Information System Database analysis compiled by the U.S. Nuclear Regulatory Commission, June 8, 2011

Nuclear Power Units by Reactor Type, Worldwide

Reactor Type	In Operation Number of Units	In Operation Net MWe
Pressurized light-water reactors (PWR)	271	249,950
Boiling light-water reactors (BWR)	88	81,367
Heavy-water reactors, all types (HWR)	47	23,042
Graphite-moderated light-water reactors (LWGR)	15	10,219
Gas-cooled reactors, all types (GCR)	18	8,949
Liquid-metal-cooled fast-breeder reactors (FBR)	1	560
Total	440	374,093

Note: MWe values rounded to the nearest whole number.

Source: IAEA Power Reactor Information System Database, www.iaea.org

Data as compiled by the U.S. Nuclear Regulatory Commission. Data available as of June 8, 2011.

APPENDIX J–K

APPENDIX L
Native American Reservations or Trust Land within a 50-Mile Radius of a Nuclear Power Plant

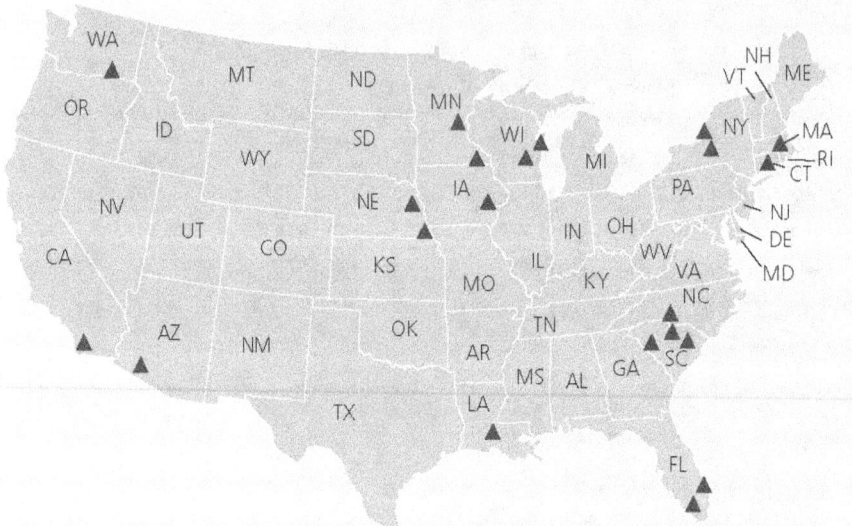

ARIZONA
Palo Verde
Ak-Chin Indian Community
Tohono O'odham
   Trust Land
Gila River Reservation
Maricopa Reserve

CALIFORNIA
San Onofre
Pechanga Reservation
   of Luiseño Indians
Pala Reservation
Pauma & Yuima Reserve
Rincon Reservation
San Pasqual Reservation
La Jolla Reservation
Cahuilla Reservation
Soboba Reservation
Santa Ysabel
Mesa Grande Reservation
Barona Reservation

CONNECTICUT
Millstone
Mohegan Reservation
Mashantucket Pequot
   Reservation
Narragansett
   Reservation

FLORIDA
St. Lucie
Brighton Reservation
   (Seminole Tribes
   of Florida)
Fort Pierce Reservation

Turkey Point
Miccosukee
   Reservation
Hollywood Reservation
   (Seminole Tribes
   of Florida)

IOWA
Duane Arnold
Sac & Fox Trust Land
Sac & Fox Reserve

LOUISIANA
River Bend
Tunica-Biloxi Reservation

MASSACHUSETTS
Pilgrim
Wampanoag
   Tribe of Grey Head
   (Aquinnah)
   Trust Land

MINNESOTA
Monticello
Shakopee Community
Shakopee Trust Land
Mille Lacs Reservation

Prairie Island
Prairie Island Community
Prairie Island Trust Land
Shakopee Community
Shakopee Trust Land

NEBRASKA
Cooper
Sac & Fox Trust Land
Sac & Fox Reservation
Kickapoo

Fort Calhoun
Winnebago Trust Land
Omaha Reservation
Winnebago Reservation

NEW YORK
FitzPatrick
Onondaga Reservation
Oneida Reservation

Nine Mile Point
Onondaga Reservation
Oneida Reservation

NORTH CAROLINA
McGuire
Catawba Reservation

SOUTH CAROLINA
Catawba
Catawba Reservation

Oconee
Eastern Cherokee
   Reservation

Summer
Catawba Reservation

WASHINGTON
Columbia
Yakama Reservation
Yakama Trust

WISCONSIN
Kewaunee
Oneida Trust Land
Oneida Reservation

Point Beach
Oneida Trust Land
Oneida Reservation

Note: This table uses NRC-abbreviated reactor names and Native American Reservation and Trust land names.

APPENDIX M
Regulatory Research Cooperative Agreements and Grants

Organization	Agreement or Grant Description
Electric Power Research Institute	Research on central and eastern United States seismic hazards, and irradiation-assisted stress-corrosion cracking
Pennsylvania State University	Assistance with a multinational research program, coordinated by the Nuclear Energy Agency, to benchmark thermal-hydraulic computer calculations against experimental data; cladding hydride reorientation and fracture behavior; and TRACE development
International Commission on Radiological Protection	Research on radiological protection standards
Oregon State University	Research on high-temperature gas reactors
University of Maryland	Research on improved human reliability analysis methods and the cause-defense approach to common cause failure modeling
University of California-Berkeley	Work on ground motion prediction models for central and eastern North America and postliquefaction residual strength
University of South Carolina	Research on aging electric cables and gas accumulation detection in nuclear power plants
University of Wisconsin	Research on advanced gas-cooled reactors
Texas A&M	Research on bypass flow in prismatic reactor blocks
American Nuclear Society	Supports the development and maintenance of probabilistic risk assessment (PRA)-related standards
ASME Standards Technology, LLC	Support in the following areas: Committee on Nuclear Risk Management on PRA standards, nuclear risk management, code comparison for the Multinational Design Evaluation Program, and a nondestructive examination certification program
National Academies	To perform a study on the cancer risk for populations surrounding nuclear power plant facilities

APPENDIX N

Issued Significant Enforcement Actions, 2010

Issued Significant Enforcement Actions, referred to as "escalated," include notices of violation for severity level (NOVSL) I, II, or III violations; notices of violation (NOV) associated with inspection findings (NOVF) that the significance determination process (SDP) categorizes as white, yellow, or red; civil penalties (CVP); and orders (CO). Escalated enforcement actions are issued to reactor, materials, and individual licensees; nonlicensees; and fuel cycle facility licensees.

Action Number	Name	Type	Issue Date	Enforcement Action
EA-09-266	Allegiance Health	Materials	1/6/2010	NOVSL III
EA-09-263	Babcock & Wilcox	Fuel Cycle Facilities	1/11/2010	NOV CVP-$35,000
EA-09-018	Entergy Operations, Inc. (Waterford Steam Electric Station)	Reactor	1/14/2010	NOV of Technical Specification 6.8.1.a, "Procedures and Programs," at Waterford Steam Electric Station, Unit 3.
EA-09-269	Entergy Nuclear Operations, Inc. (Palisades Nuclear Plant)	Reactor	1/20/2010	NOV white SDP finding result of plant inspections.
EA-09-290	Great Falls Clinic	Materials	1/21/2010	NOVSL III
EA-09-040	Chippenham/John-Willis Medical Center–Johnston-Willis Campus	Materials	1/21/2010	NOVSL III
EA-09-147	Beta Gamma Nuclear Radiology, Inc.	Materials	1/21/2010	NOV CO result of an alternative dispute resolution mediation on a violation of 10 CFR 30.9.
IA-09-041	Dr. Juan E. Perez Monté	Individual	1/21/2010	NOV CO
EA-09-248	PPL Susquehanna, LLC (Susquehanna Steam Electric Station)	Reactor	1/28/2010	NOV CO
EA-09-335	Nanticoke Memorial Hospital	Materials	2/2/2010	NOVSL III
EA-09-312	Kanawha Scales & Systems, Inc.	Materials	2/18/2010	NOVSL III
EA-09-289	Gamma Knife Center of the Pacific	Materials	2/23/2010	NOVSL III
EA-09-259	Exelon Generation Company, LLC (Braidwood Nuclear Power Station)	Reactor	2/25/2010	NOV white SDP finding result of plant inspections.
EA-09-283	FirstEnergy Nuclear Operating Company (Davis-Besse Nuclear Power Station)	Reactor	2/25/2010	NOV white SDP finding result of plant inspections.
EA-09-142	National Institute of Standards and Technology	Materials	3/1/2010	CO result of an alternate dispute resolution mediation.
IA-09-068	Lawrence Grimm	Individual	3/1/2010	CO
EA-09-082	Troxler Electronic Laboratories, Inc.	Materials	3/9/2010	NOVSL III
EA-10-014	City of South Bend, Indiana	Materials	3/10/2010	NOVSL III
EA-09-038	U.S. Department of Veterans Affairs	Materials	3/17/2010	NOVSL III–$227,000
EA-09-307	Tennessee Valley Authority (Browns Ferry Nuclear Plant)	Reactor	4/19/2010	NOV yellow and white SDP finding result of plant inspections

EA-09-321	Florida Power & Light Company (St. Lucie Nuclear Plant)	Reactor	4/19/2010	NOV yellow SDP finding
EA-10-025	SSM St. Clare Health Center	Materials	4/19/2010	NOVSL III
EA-09-272	AREVA NP, Inc.	Fuel Cycle Facilities	4/26/2010	CO
EA-09-332	FirstEnergy Nuclear Operating Company (Davis-Besse Nuclear Power Station)	Reactor	4/30/2010	NOVSL III
EA-10-009	Southern Nuclear Operating Company, Inc. (Edwin I. Hatch Nuclear Plant)	Reactor	5/12/2010	NOV white SDP finding result of plant inspections.
EA-10-063	Yale-New Haven Hospital	Materials	5/21/2010	NOVSL III
EA-09-252	Duke Energy Carolinas, LLC (William B. McGuire Nuclear Station)	Reactor	6/2/2010	CO
EA-10-044	ArcelorMittal USA, Inc.	Materials	6/2/2010	NOVSL III
EA-10-023	Department of Veterans Affairs	Materials	6/2/2010	NOVSL III CVP–$14,000
IA-09-076	Dusty Bolman	Individual	6/2/2010	NOVSL III
IA-09-075	Mary K. Files	Individual	6/2/2010	CO
EA-09-268	Global Nuclear Fuels–Americas, LLC	Fuel Cycle Facilities	6/9/2010	NOVSL III
EA-10-037	Florida Power & Light Company (Turkey Point Nuclear Plant Unit 3)	Reactor	6/21/2010	NOV white SDP finding result of plant inspections.
EA-10-068	Anthony & Edward Consultants	Materials	6/25/2010	NOVSL III
EA-10-062	Earth Engineers, Inc.	Materials	6/28/2010	NOVSL III
EA-10-069	Laboratory Testing Services, LLC	Materials	7/6/2010	NOVSL III
EA-10-110	Southern Earth Sciences, Inc.	Materials	7/19/2010	NOVSL III
EA-10-080	Calvert Cliffs Nuclear Power Plant, LLC (Calvert Cliffs Nuclear Power Plant)	Reactor	8/3/2010	NOV white SDP finding result of plant inspections.
EA-10-094	Duke Energy Carolinas, LLC (Oconee Nuclear Station)	Reactor	8/12/2010	NOV yellow and white SDP findings result of plant inspections.
EA-10-066	Bryan LGH Medical Center	Materials	8/18/2010	NOVSL III
EA-10-081	Department of Veteran Affairs	Materials	8/23/2010	2 NOVSL III CVP–$39,000
EA-10-113	Chicago Testing Laboratory, Inc.	Materials	8/24/2010	NOVSL III
EA-09-258	Basin Electric Power Cooperative	Materials	8/26/2010	NOVSL II &NOVSL III CVP–$24,000
EA-10-138	Universal Engineering Sciences, Inc.	Materials	8/27/2010	NOVSL III
EA-10-085	St. Louis Testing Laboratories, Inc.	Materials	8/31/2010	NOVSL III

IA-10-028	Mark M. Ficek	Individual	9/2/2010	CO
EA-10-086	Nuclear Fuel Services, Inc.	Fuel Cycle Facilities	9/2/2010	NOVSL III CVP–$140,000, Exercise of Enforcement Discretion
EA-10-054	Stone & Webster Construction, Inc.	Nonlicensee	9/10/2010	CO
EA-10-084	Omaha Public Power District (Fort Calhoun Station)	Reactor	10/6/2010	NOV yellow SDP finding result of plant inspections
EA-10-174	McConnell Dowell (American Samoa), Ltd.	Materials	10/6/2010	NOVSL III
EA-08-204	Babcock & Wilcox	Fuel Cycle Facilities	10/12/2010	Atomic Safety and Licensing Board—Order
EA-10-135	Analytical Bio-Chemistry Laboratories, Inc.	Materials	10/13/2010	NOVSL III
IA-10-037	Robert B. Hilton	Individual	10/20/2010	NOVSL III
EA-10-077	Superior Well Services, Ltd.	Materials	10/21/2010	NOVSL III CVP–$34,000
EA-10-140	Walter Reed Army Medical Center	Materials	10/25/2010	NOVSL III
EA-10-124	Westinghouse Electic Company	Fuel Cycle Facilities	11/3/2010	NOVSL III CVP–$17,500
EA-10-171	St. Francis Hospital and Medical Center	Materials	11/10/2010	NOVSL III
EA-10-076	Nuclear Fuel Services, Inc.	Fuel Cycle Facilities	11/16/2010	CO
EA-10-234	Kansas State University (Research Reactor Facility)	Reactor	11/22/2010	NOVSL III
EA-10-041	AREVA NP, Inc.	Materials	12/2/2010	NOV CO
IA-10-026	Richard Montgomery	Individual	12/2/2010	NOVSL III
EA-10-205	Carolina Power and Light Company (H.B. Robinson Steam Electric Plant)	Reactor	12/7/2010	NOVSL III white SDP finding result of plant inspections
EA-10-182	Sanford Medical Center	Materials	12/10/2010	NOVSL III
EA-10-207	PPL Susquehanna, LLC (Susquehanna Steam Electric Plant)	Reactor	12/16/2010	NOVSL III white SDP finding result of plant inspections
EA-10-192	Carolina Power and Light Company (Brunswick Steam Electric Plant)	Reactor	12/21/2010	NOVSL III white SDP finding result of plant inspections

Note: Reactor facilities in a decommissioning status are listed as materials licensees. The NRC report on Issued Significant Enforcement Actions can be found on the NRC Web site at www.nrc.gov/about-nrc/regulatory/enforcement/current.html.

Quick-Reference Metric Conversion Tables

SPACE AND TIME

Quantity	From Inch-Pound Units	To Metric Units	Multiply by
Length	mi (statute)	km	1.609 347
	yd	m	*0.914 4
	ft (int)	m	*0.304 8
	in	cm	*2.54
Area	mi^2	km^2	2.589 998
	acre	m^2	4 046.873
	yd^2	m^2	0.836 127 4
	ft^2	m^2	*0.092 903 04
	in^2	cm^2	*6.451 6
Volume	acre foot	m^3	1 233.489
	yd^3	m^3	0.764 554 9
	ft^3	m^3	0.028 316 85
	ft^3	L	28.316 85
	gal	L	3.785 412
	fl oz	mL	29.573 53
	in^3	cm^3	16.387 06
Velocity	mi/h	km/h	1.609 347
	ft/s	m/s	*0.304 8
Acceleration	ft/s^2	m/s^2	*0.304 8

NUCLEAR REACTION AND IONIZING RADIATION

Quantity	From Inch-Pound Units	To Metric Units	Multiply by
Activity (of a radionudide)	curie (Ci)	MBq	*37,000.0
	dpm	Becquerel (Bq)	0.016 667
Absorbed dose	rad	Gray (Gy)	*0.01
	rad	cGy	*1.0
Dose equivalent	rem	Sievert (Sv)	*0.01
	rem	mSv	*10.0
	mrem	mSv	*0.01
	mrem	µSv	*10.0
Exposure (X-rays and gamma rays)	roentgen (R)	C/kg (coulomb)	0.000 258

HEAT

Quantity	From Inch-Pound Units	To Metric Units	Multiply by
Thermodynamic temperature	°F	K	*K = (°F + 59.67)/1.8
Celsius temperature	°F	°C	*°C = (°F–32)/1.8
Linear expansion coefficient	1/°F	1/K or 1/°C	*1.8
Thermal conductivity	(Btu • in)/(ft^2 • h • °F)	W/(m • °C)	0.144 227 9
Coefficient of heat transfer	Btu / (ft^2 • h • °F)	W/(m^2 • °C)	5.678 263
Heat capacity	Btu/°F	kJ/°C	1.899 108
Specific heat capacity	Btu/(lb • °F)	kJ/(kg • °C)	*4.186 8
Entropy	Btu/°F	kJ/°C	1.899 108
Specific entropy	Btu/(lb • °F)	kJ/(kg • °C)	*4.186 8
Specific internal energy	Btu/lb	kJ/kg	*2.326

MECHANICS

Quantity	From Inch-Pound Units	To Metric Units	Multiply by
Mass (weight)	ton (short)	t (metric ton)	*0.907 184 74
	lb (avdp)	kg	*0.453 592 37
Moment of mass	lb • ft	kg • m	0.138 255
Density	ton (short)/yd^3	t/m^3	1.186 553
	lb/ft^3	g/m^3	16.018 46
Concentration (mass)	lb/gal	g/L	119.826 4
Momentum	lb • ft/s	kg • m/s	0.138 255
Angular momentum	lb • ft^2/s	kg • m^2/s	0.042 140 11
Moment of inertia	lb • ft^2	kg • m^2	0.042 140 11
Force	kip (kilopound)	kN (kilonewton)	4.448 222
	lbf	N (newton)	4.448 222
Moment of force, torque	lbf • ft	N • m	1.355 818
	lbf • in	N • m	0.122 984 8
Pressure	atm (std)	kPa (kilopascal)	*101.325
	bar	kPa	*100.0
	lbf/in^2 (formerly psi)	kPa	6.894 757
	inHg (32 °F)	kPa	3.386 38
	ftH$_2$O (39.2 °F)	kPa	2.988 98
	inH$_2$O (60 °F)	kPa	0.248 84
	mmHg (0 °C)	kPa	0.133 322

MECHANICS (continued)

Quantity	From Inch-Pound Units	To Metric Units	Multiply by
Stress	kip/in^2 (formerly ksi)	MPa	6.894 757
	lbf/in^2 (formerly psi)	MPa	0.006 894 757
	lbf/in^2 (formerly psi)	kPa	6.894 757
	lbf/ft^2	kPa	0.047 880 26
Energy, work	kWh	MJ	*3.6
	calth	J (joule)	*4.184
	Btu	kJ	1.055 056
	ft • lbf	J	1.355 818
	therm (US)	MJ	105.480 4
Power	Btu/s	kW	1.055 056
	hp (electric)	kW	*0.746
	Btu/h	W	0.293 071 1

Note: The information contained in this table is intended to familiarize NRC personnel with commonly used SI units and provide a quick reference to aid in the understanding of documents containing SI units. The conversion factors provided have not been approved as NRC guidelines for the development of licensing actions, regulations, or policy.

To convert from metric units to inch-pound units, divide the metric unit by the conversion factor.

* Exact conversion factors

Source: Federal Standard 376B (January 27, 1993), "Preferred Metric Units for General Use by the Federal Government"; and International Commission on Radiation Units and Measurements, ICRU Report 33 (1980), "Radiation Quantities and Units"

GLOSSARY (ABBREVIATIONS AND TERMS DEFINED)

Agreement State

A State that has signed an agreement with the U.S. Nuclear Regulatory Commission (NRC) authorizing the State to regulate certain uses of radioactive materials within the State.

Atomic energy

The energy that is released through a nuclear reaction or radioactive decay process. Of particular interest is the process known as fission, which occurs in a nuclear reactor and produces energy usually in the form of heat. In a nuclear power plant, this heat is used to boil water in order to produce steam that can be used to drive large turbines. This, in turn, activates generators to produce electrical power. Atomic energy is more correctly called nuclear energy.

Background radiation

The natural radiation that is always present in the environment. It includes cosmic radiation that comes from the sun and stars, terrestrial radiation that comes from the Earth, and internal radiation that exists in all living things. The typical average individual exposure in the United States from natural background sources is about 300 millirems per year.

Boiling-water reactor (BWR)

A common nuclear power reactor design in which water flows upward through the core, where it is heated by fission and allowed to boil in the reactor vessel. The resulting steam then drives turbines, which activate generators to produce electrical power. BWRs operate similarly to electrical plants using fossil fuel, except that the BWRs are powered by 370–800 nuclear fuel assemblies in the reactor core.

Brachytherapy

A nuclear medicine procedure during which a sealed radioactive source is implanted directly into a person being treated for cancer (usually of the mouth, breast, lung, prostate, ovaries, or uterus). The radioactive implant may be temporary or permanent, and the radiation attacks the tumor as long as the device remains in place. Brachytherapy uses radioisotopes, such as iridium-192 or iodine-125, which are regulated by the NRC and its Agreement States.

Byproduct material

As defined by NRC regulations includes any radioactive material (except enriched uranium or plutonium) produced by a nuclear reactor. It also includes the tailings or wastes produced by the extraction or concentration of uranium or thorium or the fabrication of fuel for nuclear reactors. Additionally, it is any material that has been made radioactive through the use of a particle accelerator or any discrete source of radium-226 used for a commercial, medical, or research activity. In addition, the NRC, in consultation with the U.S. Environmental Protection Agency (EPA), U.S. Department of Energy (DOE), U.S. Department of Homeland Security (DHS), and others, can designate as byproduct material any source of naturally-occurring radioactive material, other than source material, that it determines would pose a threat to public health and safety or the common defense and security of the United States.

GLOSSARY

Canister

See *Dry cask storage*.

Capability

The maximum load that a generating unit, generating station, or other electrical apparatus can carry under specified conditions for a given period of time without exceeding approved limits of temperature and stress.

Capacity

The amount of electric power that a generating unit can produce. The amount of electric power that a manufacturer rates its generator, turbine transformer, transmission, circuit, or system as able to produce.

Capacity charge

One of two elements in a two-part pricing method used in capacity transactions (the other element is the energy charge). The capacity charge, sometimes called the demand charge, is assessed on the capacity (amount of electric power) being purchased.

Capacity factor

The ratio of the available capacity (the amount of electrical power actually produced by a generating unit) to the theoretical capacity (the amount of electrical power that could theoretically have been produced if the generating unit had operated continuously at full power) during a given time period.

Capacity utilization

A percentage representing the extent to which a generating unit fulfilled its capacity in generating electric power over a given time period. This percentage is defined as the margin between the unit's available capacity (the amount of electrical power the unit actually produced) and its theoretical capacity (the amount of electrical power that could have been produced if the unit had operated continuously at full power) during a certain time period. Capacity utilization is computed by dividing the amount actually produced by the theoretical capacity, and multiplying by 100.

Cask

A heavily shielded container used for the dry storage or shipment (or both) of radioactive materials such as spent nuclear fuel or other high-level radioactive waste (HLW). Casks are often made from lead, concrete, or steel. Casks must meet regulatory requirements and are not intended for long-term disposal in a repository.

Classified information

Information that could be used by an adversary to harm the United States or its allies and thus must be protected. The NRC has two types of classified information. The first type, known as national security information, is information that is classified by an Executive order. Its release would damage national security to some degree. The second type, known as restricted data, is information that is classified by the Atomic Energy Act of 1954, as amended. It would assist individuals or organizations in designing, manufacturing, or using nuclear weapons. Access to both types of information is restricted to authorized persons who have been properly cleared and have a "need to know" the information for their official duties.

Combined license (COL)

An NRC-issued license that authorizes a licensee to construct and (with certain specified conditions) operate a nuclear power plant at a specific site, in accordance with established laws and regulations. A COL is valid for 40 years (with the possibility of a 20-year renewal).

Commercial sector (energy users)

Generally, nonmanufacturing business establishments, including hotels, motels, and restaurants; wholesalers and retail stores; and health, social, and educational institutions. However, utilities may categorize commercial service as all consumers whose demand or annual usage exceeds some specified limit that is categorized as residential service.

Compact

A group of two or more States that have formed business alliances to dispose of low-level radioactive waste (LLW) on a regional basis.

Construction recapture

The maximum number of years that could be added to a facility's license expiration date to recapture the period between the date the NRC issued the facility's construction permit to the date it granted an operating license. A licensee must submit an application to request this extension.

Containment structure

A gas-tight shell or other enclosure around a nuclear reactor to confine fission products that otherwise might be released to the atmosphere in the event of an accident. Such enclosures are usually dome-shaped and made of steel-reinforced concrete.

Contamination

Undesirable radiological, chemical, or biological material (with a potentially harmful effect) that is either airborne or deposited in (or on the surface of) structures, objects, soil, water, or living organisms in a concentration that makes the medium unfit for its next intended use.

Criticality

The normal operating condition of a reactor, in which nuclear fuel sustains a fission chain reaction. A reactor achieves criticality (and is said to be critical) when each fission event releases a sufficient number of neutrons to sustain an ongoing series of reactions.

Decommissioning

The process of safely closing a nuclear power plant (or other facility where nuclear materials are handled) to retire it from service after its useful life has ended. This process primarily involves decontaminating the facility to reduce residual radioactivity and then releasing the property for unrestricted or (under certain conditions) restricted use. This often includes dismantling the facility or dedicating it to other purposes. Decommissioning begins after the nuclear fuel, coolant, and radioactive waste are removed.

GLOSSARY

DECON

A method of decommissioning, in which structures, systems, and components that contain radioactive contamination are removed from a site and safely disposed at a commercially operated low-level waste disposal facility, or decontaminated to a level that permits the site to be released for unrestricted use shortly after it ceases operation.

Decontamination

A process used to reduce, remove, or neutralize radiological, chemical, or biological contamination to reduce the risk of exposure. Decontamination may be accomplished by cleaning or treating surfaces to reduce or remove the contamination; filtering contaminated air or water; subjecting contamination to evaporation and precipitation; or covering the contamination to shield or absorb the radiation. The process can also simply allow adequate time for natural radioactive decay to decrease the radioactivity.

Defense-in-depth

An approach to designing and operating nuclear facilities that prevents and mitigates accidents that release radiation or hazardous materials. The key is creating multiple independent and redundant layers of defense to compensate for potential human and mechanical failures so that no single layer, no matter how robust, is exclusively relied upon. Defense-in-depth includes the use of access controls, physical barriers, redundant and diverse key safety functions, and emergency response measures.

Depleted uranium

Uranium with a percentage of uranium-235 lower than the 0.7 percent (by mass) contained in natural uranium. (The normal residual uranium-235 content in depleted uranium is 0.2–0.3 percent, with uranium-238 comprising the remaining 98.7–98.8 percent.) Depleted uranium is produced during uranium isotope separation and is typically found in spent fuel elements or byproduct tailings or residues. Depleted uranium can be blended with highly-enriched uranium, such as that from weapons, to make reactor fuel.

Design-basis threat (DBT)

A profile of the type, composition, and capabilities of an adversary. The NRC uses the DBT as a basis for designing safeguards systems to protect against acts of radiological sabotage and to prevent the theft of special nuclear material. Nuclear facility licensees are expected to demonstrate they can defend against the DBT.

Design certification

Certification and approval by the NRC of a standard nuclear power plant design independent of a specific site or an application to construct or operate a plant. A design certification is valid for 15 years from the date of issuance but can be renewed for an additional 10 to 15 years.

Dry cask storage

A method for storing spent nuclear fuel above ground in special containers known as casks. After fuel has been cooled in a spent fuel pool for at least 1 year, dry cask storage allows approximately one to six dozen spent fuel assemblies to be sealed in casks and surrounded by inert gas. The casks are large, rugged cylinders made of steel or steel-reinforced concrete (18 or more inches thick or 45.72 or more centimeters).

They are welded or bolted closed, and each cask is surrounded by steel, concrete, lead, or other material to provide leak-tight containment and radiation shielding. The casks may be placed horizontally in aboveground concrete bunkers, or vertically in concrete vaults or on concrete pads.

Early site permit (ESP)

A permit through which the NRC resolves site safety, environmental protection, and emergency preparedness (EP) issues, in order to approve one or more proposed sites for a nuclear power facility, independent of a specific nuclear plant design or an application for a construction permit or COL. An ESP is valid for 10 to 20 years but can be renewed for an additional 10 to 20 years.

Economic Simplified Boiling-Water Reactor (ESBWR)

A 4,500-megawatts thermal nuclear reactor design, which has passive safety features and uses natural circulation (with no recirculation pumps or associated piping) for normal operation. GE-Hitachi Nuclear Energy (GEH) submitted an application for final design approval and standard design certification for the ESBWR on August 24, 2005.

Efficiency, plant

The percentage of the total energy content of a power plant's fuel that is converted into electricity. The remaining energy is lost to the environment as heat.

Electric power grid

A system of synchronized power providers and consumers, connected by transmission and distribution lines and operated by one or more control centers. In the continental United States, the electric power grid consists of three systems—the Eastern Interconnect, the Western Interconnect, and the Texas Interconnect. In Alaska and Hawaii, several systems encompass areas smaller than the State.

Electric utility

A corporation, agency, authority, person, or other legal entity that owns and/or operates facilities within the United States, its territories, or Puerto Rico for the generation, transmission, distribution, or sale of electric power (primarily for use by the public). Facilities that qualify as cogenerators or small power producers under the Public Utility Regulatory Policies Act are not considered electric utilities.

Emergency classifications

Sets of plant conditions that indicate various levels of risk to the public and that might require response by an offsite emergency response organization to protect citizens near the site.

Emergency preparedness (EP)

The programs, plans, training, exercises, and resources necessary to prepare emergency personnel to rapidly identify, evaluate, and react to emergencies, including those arising from terrorism or natural events such as hurricanes. EP strives to ensure that nuclear power plant operators can implement measures to protect public health and safety in the event of a radiological emergency. Plant operators, as a condition of their licenses, must develop and maintain EP plans that meet NRC requirements.

Energy Information Administration (EIA)

The agency, within the U.S. Department of Energy, that provides policy-neutral statistical data, forecasts, and analyses to promote sound policymaking, efficient markets, and public understanding regarding energy and its interaction with the economy and the environment.

ENTOMB

A method of decommissioning, in which radioactive contaminants are encased in a structurally long-lived material, such as concrete. The entombed structure is maintained and surveillance is continued until the entombed radioactive waste decays to a level permitting termination of the license and unrestricted release of the property. During the entombment period, the licensee maintains the license previously issued by the NRC.

Event Notification System

An automated event tracking system used internally by the NRC's Headquarters Operations Center to track incoming notifications of significant nuclear events with an actual or potential effect on the health and safety of the public and the environment. Significant events are reported to the Operations Center by the NRC's licensees, Agreement States, other Federal agencies, the public, and other stakeholders.

Exposure

Absorption of ionizing radiation or ingestion of a radioisotope. Acute exposure is a large exposure received over a short period of time. Chronic exposure is exposure received over a long period of time, such as during a lifetime. The National Council on Radiation Protection and Measurements (NCRP) estimates that an average person in the United States receives a total annual dose of about 0.62 rem (620 millirem) from all radiation source, a level that has not been shown to cause humans any harm. Of this total, natural background sources of radiation—including radon and thoron gas, natural radiation from soil and rocks, radiation from space, and radiation sources that are found naturally within the human body—account for approximately 50 percent. Medical procedures such as computed tomography (CT scans) and nuclear medicine account approximately for another 48 percent. Other small contributors of exposure to

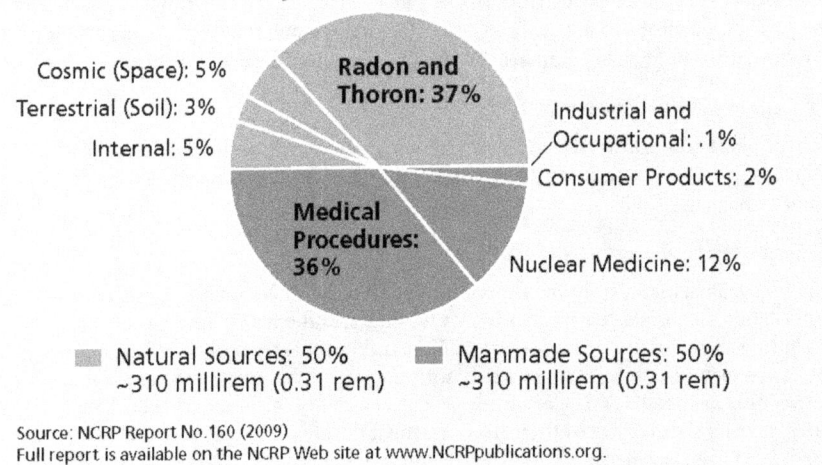

Sources of Radiation Exposure in the United States

Cosmic (Space): 5%
Terrestrial (Soil): 3%
Internal: 5%
Radon and Thoron: 37%
Industrial and Occupational: .1%
Consumer Products: 2%
Medical Procedures: 36%
Nuclear Medicine: 12%

Natural Sources: 50%
~310 millirem (0.31 rem)

Manmade Sources: 50%
~310 millirem (0.31 rem)

Source: NCRP Report No.160 (2009)
Full report is available on the NCRP Web site at www.NCRPpublications.org.

the U.S. population include consumer products and activities, industrial and research uses, and occupational tasks. The maximum permissible yearly dose for a person working with or around nuclear material is 5 rem.

Federal Emergency Management Agency (FEMA)

A component of DHS responsible for protecting the nation and reducing the loss of life and property from all hazards, such as natural disasters and acts of terrorism. FEMA leads and supports a risk-based, comprehensive emergency management system of preparedness, protection, response, recovery, and mitigation. FEMA also administers the National Flood Insurance Program.

Federal Energy Regulatory Commission (FERC)

An independent agency that regulates the interstate transmission of electricity, natural gas, and oil. FERC also regulates and oversees hydropower projects and the construction of liquefied natural gas terminals and interstate natural gas pipelines. FERC protects the economic, environmental, and safety interests of the American public, while working to assure abundant, reliable energy in a fair, competitive market.

Fiscal year (FY)

The 12-month period from October 1 through September 30 used by the Federal Government for budget formulation and execution. The FY is designated by the calendar year in which it ends; for example, FY 2009 runs from October 1, 2008, through September 30, 2009.

Fissile material

A nuclide that is capable of undergoing fission after capturing low-energy thermal (slow) neutrons. Although sometimes used as a synonym for fissionable material, this term has acquired its more restrictive interpretation with the limitation that the nuclide must be fissionable by thermal neutrons. With that interpretation, the three primary fissile materials are uranium-233, uranium-235, and plutonium-239. This definition excludes natural uranium and depleted uranium that have not been irradiated or have only been irradiated in thermal reactors.

Fission (fissioning)

The splitting of an atom, which releases a considerable amount of energy (usually in the form of heat) that can be used to produce electricity. Fission may be spontaneous but is usually caused by the nucleus of an atom becoming unstable (or "heavy") after capturing or absorbing a neutron. During fission, the heavy nucleus splits into roughly equal parts, producing the nuclei of at least two lighter elements. In addition to energy, this reaction usually releases gamma radiation and two or more daughter neutrons.

Force-on-force

Inspections designed to evaluate and improve the effectiveness of a licensee's security force and ability to defend a nuclear power plant and other nuclear facilities against a DBT. An essential part of the security program instituted by the NRC, a full force-on-force inspection spans 2 weeks and includes tabletop drills and multiple simulated combat exercises between a mock commando-type adversary force and the plant's security force.

GLOSSARY

Foreign Assignee Program

An on-the-job training program, sponsored by the NRC for assignees from other countries, usually under bilateral information exchange arrangements with their respective regulatory organizations.

Freedom of Information Act (FOIA)

A Federal law that requires Federal agencies to provide, upon written request, access to records or information. Some material is exempt from FOIA, and FOIA does not apply to records that are maintained by State and local governments, or Federal contractors, grantees, or private organizations or businesses.

Fuel assembly (fuel bundle, fuel element)

A structured group of fuel rods (long, slender, metal tubes containing pellets of fissionable material, which provide fuel for nuclear reactors). Depending on the design, each reactor vessel may have dozens of fuel assemblies (also known as fuel bundles), each of which may contain 200 or more fuel rods.

Fuel cycle

The series of steps involved in supplying fuel for nuclear power reactors includes the following:

- Uranium recovery to extract (or mine) uranium ore and concentrate (or mill) the ore to produce "yellowcake"

- Conversion of yellowcake into uranium hexafluoride (UF_6)

- Enrichment to increase the concentration of uranium-235 in UF_6

- Fuel fabrication to convert enriched UF_6 into fuel for nuclear reactors

- Use of the fuel in reactors (nuclear power, research, or naval propulsion)

- Interim storage of spent nuclear fuel

- Reprocessing of high-level waste to recover the fissionable material remaining in the spent fuel (currently not done in the United States)

- Final disposition (disposal) of high-level waste

The NRC regulates these processes, as well as the fabrication of mixed oxide (MOX) nuclear fuel, which is a combination of uranium and plutonium oxides.

Fuel reprocessing (recycling)

The processing of reactor fuel to separate the unused fissionable material from waste material. Reprocessing extracts isotopes from spent nuclear fuel so they can be used again as reactor fuel. Commercial reprocessing is not practiced in the United States, although it has been practiced in the past. However, the U.S. Department of Defense oversees reprocessing programs at DOE facilities such as in Hanford, WA, and Savannah River, SC. These wastes as well as those wastes at a formerly operating commercial reprocessing facility at West Valley, NY, are not regulated by the NRC.

Fuel rod

A long, slender, zirconium metal tube containing pellets of fissionable material, which provide fuel for nuclear reactors. Fuel rods are assembled into bundles called fuel assemblies, which are loaded individually into the reactor core.

Full-time equivalent (FTE)

A human resources measurement equal to one staff person working full-time for one year.

Gas centrifuge

A uranium enrichment process used to prepare uranium for use in fabricating fuel for nuclear reactors by separating its isotopes (as gases) based on their slight difference in mass. This process uses a large number of interconnected centrifuge machines (rapidly spinning cylinders). No commercial gas centrifuge plants are operating in the United States; however, both Louisiana Energy Services and United States Enrichment Corporation (USEC) have received licenses to construct and operate such facilities, and both facilities are under construction.

Gas chromatography

A way of separating chemical substances from a mixed sample by passing the sample, carried by a moving stream of gas, through a tube packed with a finely divided solid that may be coated with a liquid film. Gas chromatography devices are used to analyze air pollutants, blood alcohol content, essential oils, and food products.

Gaseous diffusion

A uranium enrichment process used to prepare uranium for use in fabricating fuel for nuclear reactors by separating its isotopes (as gases) based on their slight difference in velocity. (Lighter isotopes diffuse faster through a porous membrane or vessel than do heavier isotopes.) This process involves filtering UF_6 gas to separate uranium-234 and uranium-235 from uranium-238, in order to increase the percentage of uranium-235 from 1 to 3 percent. The only gaseous diffusion plant in operation in the United States is in Paducah, KY. A similar plant near Piketon, OH, was closed in March 2001. Both plants are leased by the USEC from DOE and have been regulated by the NRC since March 4, 1997.

Gauging devices

Devices used to measure, monitor, and control the thickness of sheet metal, textiles, paper napkins, newspaper, plastics, photographic film, and other products as they are manufactured. Gauges mounted in fixed locations are designed for measuring or controlling material density, flow, level, thickness, or weight. The gauges contain sealed sources that radiate through the substance being measured to a readout or controlling device. Portable gauging devices, such as moisture density gauges, are used at field locations. These gauges contain a gamma-emitting sealed source, usually cesium-137, or a sealed neutron source, usually americium-241 or beryllium.

Generation (gross)

The total amount of electric energy produced by a generating station, as measured at the generator terminals.

GLOSSARY

Generation (net)

The gross amount of electric energy produced by a generating station, minus the amount used to operate the station. Net generation is usually measured in watthours.

Generator capacity

The maximum amount of electric energy that a generator can produce (from the mechanical energy of the turbine), adjusted for ambient conditions. Generator capacity is commonly expressed in megawatts (MW).

Generator nameplate capacity

The maximum amount of electric energy that a generator can produce under specific conditions, as rated by the manufacturer. Generator nameplate capacity is usually expressed in kilovolt-amperes and kilowatts (kW), as indicated on a nameplate that is physically attached to the generator.

Geological repository

An excavated, underground facility that is designed, constructed, and operated for safe and secure permanent disposal of HLW. A geological repository uses an engineered barrier system and a portion of the site's natural geology, hydrology, and geochemical systems to isolate the radioactivity of the waste. The Nuclear Waste Policy Act of 1982, as amended, specifies that this waste will be disposed of in a deep geologic repository, and that Yucca Mountain, NV, will be the single candidate site for such a repository. On June 3, 2008, DOE submitted a license application to the NRC seeking authorization to construct the Yucca Mountain repository. On January 29, 2010, the President created the Blue Ribbon Commission on America's Nuclear Future to reassess the national policy on high-level waste disposal.

Gigawatt (GW)

A unit of power equivalent to one billion watts.

Gigawatthour (GWh)

One billion watthours.

Grid

See *Electric power grid*.

Half-life (radiological)

The time required for half the atoms of a particular radioisotope to decay into another isotope that has half the activity of the original radioisotope. A specific half-life is a characteristic property of each radioisotope. Measured half-lives range from millionths of a second to billions of years, depending on the stability of the nucleus. Radiological half-life is related to, but different from, the biological half-life and the effective half-life.

Health physics

The science concerned with recognizing and evaluating the effects of ionizing radiation on the health and safety of people and the environment, monitoring radiation exposure, and controlling the associated health risks and environmental hazards to permit the safe use of technologies that produce ionizing radiation.

High-level radioactive waste (HLW)

The highly radioactive materials produced as byproducts of fuel reprocessing or of the reactions that occur inside nuclear reactors. HLW includes the following:

* irradiated spent nuclear fuel discharged from commercial nuclear power reactors

* the highly radioactive liquid and solid materials resulting from the reprocessing of spent nuclear fuel, which contain fission products in concentration (this includes some reprocessed HLW from defense activities and a small quantity of reprocessed commercial HLW)

* other highly radioactive materials that the Commission may determine require permanent isolation

Highly (or High-) enriched uranium

Uranium enriched to at least 20 percent uranium-235 (a higher concentration than exists in natural uranium ore).

In situ recovery (ISR)

One of the two primary recovery methods that are currently used to extract uranium from ore bodies where they are normally found underground (in other words, in situ), without physical excavation. Also known as "solution mining" or in situ leaching.

Incident response

Activities that address the short-term, direct effects of a natural or human-caused event and require an emergency response to protect life or property.

Independent spent fuel storage installation (ISFSI)

A complex designed and constructed for the interim storage of spent nuclear fuel; solid, reactor-related, greater than Class C waste; and other associated radioactive materials. A spent fuel storage facility may be considered independent, even if it is located on the site of another NRC-licensed facility.

International Atomic Energy Agency (IAEA)

The center of worldwide cooperation in the nuclear field, through which member countries and multiple international partners work together to promote the safe, secure, and peaceful use of nuclear technologies. The United Nations established IAEA in 1957 as "Atoms for Peace."

International Nuclear Regulators Association

An association established in January 1997 to give international nuclear regulators a forum to discuss nuclear safety. Countries represented include Canada, France, Japan, Spain, South Korea, Sweden, the United Kingdom, and the United States.

Irradiation

Exposure to ionizing radiation. Irradiation may be intentional, such as in cancer treatments or in sterilizing medical instruments. Irradiation may also be accidental, such as being exposed to an unshielded source. Irradiation does not usually result in radioactive contamination, but damage can occur, depending on the dose received.

GLOSSARY

Isotope

Two or more forms (or atomic configurations) of a given element that have identical atomic numbers (the same number of protons in their nuclei) and the same or very similar chemical properties but different atomic masses (different numbers of neutrons in their nuclei) and distinct physical properties. Thus, carbon-12, carbon-13, and carbon-14 are isotopes of the element carbon, and the numbers denote the approximate atomic masses. Among their distinct physical properties, some isotopes (known as radioisotopes) are radioactive because their nuclei emit radiation as they strive toward a more stable nuclear configuration. For example, carbon-12 and carbon-13 are stable, but carbon-14 is unstable and radioactive.

Kilowatt (kW)

A unit of power equivalent to one thousand watts.

Licensed material

Source material, byproduct material, or special nuclear material that is received, possessed, used, transferred, or disposed of under a general or specific license issued by the NRC or Agreement States.

Licensee

A company, organization, institution, or other entity to which the NRC has granted a general or specific license to construct or operate a nuclear facility, or to receive, possess, use, transfer, or dispose of source, byproduct, or special nuclear material.

Licensing basis

The collection of documents or technical criteria that provides the basis upon which the NRC issues a license to construct or operate a nuclear facility; to conduct operations involving the emission of radiation; or to receive, possess, use, transfer, or dispose of source, byproduct, or special nuclear material.

Light-water reactor

A term used to describe reactors using ordinary water as a coolant, including BWRs and PWRs, the most common types used in the United States.

Low-level radioactive waste (LLW)

A general term for a wide range of items that have become contaminated with radioactive material or have become radioactive through exposure to neutron radiation. A variety of industries, hospitals and medical institutions, educational and research institutions, private or government laboratories, and nuclear fuel cycle facilities generate LLW as part of their day-to-day use of radioactive materials. Some examples include radioactively contaminated protective shoe covers and clothing; cleaning rags, mops, filters, and reactor water treatment residues; equipment and tools; medical tubes, swabs, and hypodermic syringes; and carcasses and tissues from laboratory animals. The radioactivity in these wastes can range from just above natural background levels to much higher levels, such as seen in parts from inside the reactor vessel in a nuclear power plant. Low-level waste is typically stored onsite by licensees, either until it has decayed away and can be disposed of as ordinary trash, or until the accumulated amount becomes large enough to warrant shipment to a low-level waste disposal site.

Maximum dependable capacity (gross)

The maximum amount of electricity that the main generating unit of a nuclear power reactor can reliably produce during the summer or winter (usually summer, but whichever represents the most restrictive seasonal conditions, with the least electrical output). The dependable capacity varies during the year because temperature variations in cooling water affect the unit's efficiency. Thus, this is the gross electrical output as measured (in watts unless otherwise noted) at the output terminals of the turbine generator.

Maximum dependable capacity (net)

The gross maximum dependable capacity of the main generating unit in a nuclear power reactor, minus the amount used to operate the station. Net maximum dependable capacity is measured in watts unless otherwise noted.

Megawatt (MW)

A unit of power equivalent to one million watts.

Metric ton

Approximately 2,200 pounds.

Mill tailings

Primarily, the sandy process waste material from a conventional uranium recovery facility. This naturally radioactive ore residue contains the radioactive decay products from the uranium chains (mainly the uranium-238 chain) and heavy metals. Although the milling process recovers about 93 percent of the uranium, the residues (known as "tailings") contain several naturally occurring radioactive elements, including uranium, thorium, radium, polonium, and radon.

Mixed oxide (MOX) fuel

A type of nuclear reactor fuel (often called "MOX") that contains plutonium oxide mixed with either natural or depleted uranium oxide, in ceramic pellet form. (This differs from conventional nuclear fuel, which is made of pure uranium oxide.) Using plutonium reduces the amount of highly enriched uranium needed to produce a controlled reaction in commercial light-water reactors. However, plutonium exists only in trace amounts in nature and, therefore, must be produced by neutron irradiation of uranium-238 or obtained from other manufactured sources. As directed by Congress, the NRC regulates the fabrication of MOX fuel by DOE, a program that is intended to dispose of plutonium from international nuclear disarmament agreements.

Monitoring of radiation

Periodic or continuous determination of the amount of ionizing radiation or radioactive contamination in a region. Radiation monitoring is a safety measure to protect the health and safety of the public and the environment through the use of bioassay, alpha scans, and other radiological survey methods to monitor air, surface water and ground water, soil and sediment, equipment surfaces, and personnel.

GLOSSARY

National Response Framework

The guiding principles, roles, and structures that enable all domestic incident response partners to prepare for and provide a unified national response to disasters and emergencies. It describes how the Federal Government, States, Tribes, communities, and the private sector work together to coordinate a national response. The framework, which became effective March 22, 2008, builds upon the National Incident Management System, which provides a template for managing incidents.

National Source Tracking System (NSTS)

A secure, Web-based data system that helps the NRC and its Agreement States track and regulate the medical, industrial, and academic uses of certain nuclear materials, from the time they are manufactured or imported to the time of their disposal or exportation. This information enhances the ability of the NRC and Agreement States to conduct inspections and investigations, communicate information to other government agencies, and verify the ownership and use of nationally tracked sources.

Natural uranium

Uranium containing the relative concentrations of isotopes found in nature (0.7 percent uranium-235, 99.3 percent uranium-238, and a trace amount of uranium-234 by mass). In terms of radioactivity, however, natural uranium contains approximately 2.2 percent uranium-235, 48.6 percent uranium-238, and 49.2 percent uranium-234. Natural uranium can be used as fuel in nuclear reactors.

Net electric generation

The gross amount of electric energy produced by a generating station, minus the amount used to operate the station. Note: Electricity required for pumping at pumped-storage plants is regarded as electricity for station operation and is deducted from gross generation. Net electric generation is measured in watthours, except as otherwise noted.

Net summer capacity

The steady hourly output that generating equipment is expected to supply to system load, exclusive of auxiliary power, as demonstrated by measurements at the time of peak demand (summer). Net summer capacity is measured in watts unless otherwise noted.

Nonpower reactor (research and test reactor)

A nuclear reactor that is used for research, training, or development purposes (which may include producing radioisotopes for medical and industrial uses) but has no role in producing electrical power. These reactors, which are also known as research and test reactors, contribute to almost every field of science, including physics, chemistry, biology, medicine, geology, archeology, and ecology.

NRC Operations Center

The primary center of communication and coordination among the NRC, its licensees, State and Tribal agencies, and other Federal agencies regarding operating events involving nuclear reactors or materials. Located in Rockville, MD, the Operations Center is staffed 24 hours a day by employees trained to receive and evaluate event reports and coordinate incident response activities.

Nuclear energy

See *Atomic energy*.

Nuclear Energy Agency (NEA)

A specialized agency within the Organisation for Economic Co-operation and Development (OECD), which was created to assist its member countries in maintaining and further developing the scientific, technological, and legal bases for safe, environmentally friendly, and economical use of nuclear energy for peaceful purposes. The NEA's current membership consists of 28 countries in Europe, North America, and the Asia-Pacific region, which account for approximately 85 percent of the world's installed nuclear capacity.

Nuclear fuel

Fissionable material that has been enriched to a composition that will support a self-sustaining fission chain reaction when used to fuel a nuclear reactor, thereby producing energy (usually in the form of heat or useful radiation) for use in other processes.

Nuclear materials

See *Special nuclear material*, *Source material*, and *Byproduct material*.

Nuclear Material Management and Safeguards System (NMMSS)

A centralized U.S. Government database used to track and account for source and special nuclear material, to ensure that it has not been stolen or diverted to unauthorized users. The system contains current and historical data on the possession, use, and shipment of source and special nuclear material within the United States, as well as all exports and imports of such material. The database is jointly funded by the NRC and DOE and is operated under a DOE contract.

Nuclear poison (or neutron poison)

In reactor physics, a substance (other than fissionable material) that has a large capacity for absorbing neutrons in the vicinity of the reactor core. This effect may be undesirable in some reactor applications because it may prevent or disrupt the fission chain reaction, thereby affecting normal operation. However, neutron-absorbing materials (commonly known as "poisons") are intentionally inserted into some types of reactors to decrease the reactivity of their initial fresh fuel load. (Adding poisons, such as control rods or boron, is described as adding "negative reactivity" to the reactor.)

Nuclear power plant

A thermal power plant, in which the energy (heat) released by the fissioning of nuclear fuel is used to boil water to produce steam. The steam spins the propeller-like blades of a turbine that turns the shaft of a generator to produce electricity. Of the various nuclear power plant designs, PWRs and BWRs are in commercial operation in the United States. These facilities generate about 20 percent of U.S. electrical power.

Nuclear/Radiological Incident Annex

An annex to the National Response Framework, which provides for a timely, coordinated response by Federal agencies to nuclear or radiological accidents or incidents within the United States. This annex covers radiological dispersal devices and improvised nuclear devices, as well as accidents involving commercial reactors or weapons production facilities, lost radioactive sources, transportation accidents involving radioactive material, and foreign accidents involving nuclear or radioactive material.

GLOSSARY

Nuclear reactor

The heart of a nuclear power plant or nonpower reactor, in which nuclear fission may be initiated and controlled in a self-sustaining chain reaction to generate energy or produce useful radiation. Although there are many types of nuclear reactors, they all incorporate certain essential features, including the use of fissionable material as fuel, a moderator (such as water) to increase the likelihood of fission (unless reactor operation relies on fast neutrons), a reflector to conserve escaping neutrons, coolant provisions for heat removal, instruments for monitoring and controlling reactor operation, and protective devices (such as control rods and shielding).

Nuclear waste

A subset of radioactive waste that includes unusable byproducts produced during the various stages of the nuclear fuel cycle, including extraction, conversion, and enrichment of uranium; fuel fabrication; and use of the fuel in nuclear reactors. Specifically, these stages produce a variety of nuclear waste materials, including uranium mill tailings, depleted uranium, and spent (depleted) fuel, all of which are regulated by the NRC. (By contrast, "radioactive waste" is a broader term, which includes all wastes that contain radioactivity, regardless of how they are produced. It is not considered "nuclear waste" because it is not produced through the nuclear fuel cycle and is generally not regulated by the NRC.)

Occupational dose

The internal and external dose of ionizing radiation received by workers in the course of employment in such areas as fuel cycle facilities, industrial radiography, nuclear medicine, and nuclear power plants. These workers are exposed to varying amounts of radiation, depending on their jobs and the sources with which they work. The NRC requires its licensees to limit occupational exposure to 5,000 mrem (50 millisievert) per year. Occupational dose does not include the dose received from natural background sources, doses received as a medical patient or participant in medical research programs, or "second-hand doses" received through exposure to individuals treated with radioactive materials.

Organisation for Economic Co-operation and Development (OECD)

An intergovernmental organization (based in Paris, France) which provides a forum for discussion and cooperation among the governments of industrialized countries committed to democracy and the market economy. The primary goal of the OECD and its member countries is to support sustainable economic growth, boost employment, raise living standards, maintain financial stability, assist other countries' economic development, and contribute to growth in world trade. In addition, the OECD is a reliable source of comparable statistics and economic and social data. The OECD also monitors trends, analyzes and forecasts economic developments, and researches social changes and evolving patterns in trade, environment, agriculture, technology, taxation, and other areas.

Orphan sources (unwanted radioactive material)

Sealed sources of radioactive material contained in a small volume (but not radioactively contaminated soils and bulk metals) in any one or more of the following conditions:

- an uncontrolled condition that requires removal to protect public health and safety from a radiological threat

- a controlled or uncontrolled condition, for which a responsible party cannot be readily identified

- a controlled condition, compromised by an inability to ensure the continued safety of the material (e.g., the licensee may have few or no options to provide for safe disposition of the material)

- an uncontrolled condition, in which the material is in the possession of a person who did not seek, and is not licensed, to possess it

- an uncontrolled condition, in which the material is in the possession of a State radiological protection program solely to mitigate a radiological threat resulting from one of the above conditions, and for which the State does not have the necessary means to provide for the appropriate disposition of the material

Outage

The period during which a generating unit, transmission line, or other facility is out of service. Outages may be forced or scheduled, and full or partial.

Outage (forced)

The shutdown of a generating unit, transmission line, or other facility for emergency reasons, or a condition in which the equipment is unavailable as a result of an unanticipated breakdown. An outage (whether full, partial, or attributable to a failed start) is considered "forced" if it could not reasonably be delayed beyond 48 hours from identification of the problem, if there had been a strong commercial desire to do so. In particular, the following problems may result in forced outages:

- any failure of mechanical, fuel handling, or electrical equipment or controls within the generator's ownership or direct responsibility (i.e., from the point the generator is responsible for the fuel through to the electrical connection point)

- a failure of a mine or fuel transport system dedicated to that power station with a resulting fuel shortage that cannot be economically managed

- inadvertent or operator error

- limitations caused by fuel quality

Forced outages do not include scheduled outages for inspection, maintenance, or refueling.

Outage (full forced)

A forced outage that causes a generating unit to be removed from the Committed state (when the unit is electrically connected and generating or pumping) or the Available state (when the unit is available for dispatch as a generator or pump but is not electrically connected and not generating or pumping). Full-forced outages do not include failed starts.

GLOSSARY

Outage (scheduled)

The shutdown of a generating unit, transmission line, or other facility for inspection, maintenance, or refueling, which is scheduled well in advance (even if the schedule changes). Scheduled outages do not include forced outages and could be deferred if there were a strong commercial reason to do so.

Pellet, fuel

A thimble-sized ceramic cylinder (approximately 3/8-inch in diameter and 5/8-inch in length), consisting of uranium (typically uranium oxide), which has been enriched to increase the concentration of uranium-235 (U-235) to fuel a nuclear reactor. Modern reactor cores in PWRs and BWRs may contain up to 10 million pellets, stacked in the fuel rods that form fuel assemblies.

Performance-based regulation

A regulatory approach that focuses on desired, measurable outcomes, rather than prescriptive processes, techniques, or procedures. Performance-based regulation leads to defined results without specific direction regarding how those results are to be obtained. At the NRC, performance-based regulatory actions focus on identifying performance measures that ensure an adequate safety margin and offer incentives for licensees to improve safety without formal regulatory intervention by the agency.

Performance indicator

A quantitative measure of a particular attribute of licensee performance that shows how well a plant is performing when measured against established thresholds. Licensees submit their data quarterly; the NRC regularly conducts inspections to verify the submittals and then uses its own inspection data plus the licensees' submittals to assess each plant's performance.

Possession-only license

A license, issued by the NRC, that authorizes the licensee to possess specific nuclear material but does not authorize its use or the operation of a nuclear facility.

Power uprate

The process of increasing the maximum power level a commercial nuclear power plant may operate. This power level, regulated by the NRC, is included in the plant's operating license and technical specifications. A licensee may only change its maximum power output after the NRC approves an uprate application. The NRC analyses must demonstrate that the plant could continue to operate safely with its proposed new configuration. When all requisite conditions are fulfilled, the NRC may grant the power uprate by amending the plant's operating license and technical specifications.

Pressurized-water reactor (PWR)

A common nuclear power reactor design in which very pure water is heated to a very high temperature by fission, kept under high pressure (to prevent it from boiling), and converted to steam by a steam generator (rather than by boiling, as in a BWR). The resulting steam is used to drive turbines, which activate generators to produce electrical power. A PWR essentially operates like a pressure cooker, where a lid is tightly placed over a pot of heated water, causing the pressure inside to increase as the temperature increases (because the steam cannot escape) but keeping the water from boiling at the usual 212 degrees Fahrenheit (100 degrees Celsius). About two-thirds of the operating nuclear reactor power plants in the United States are PWRs.

Probabilistic risk assessment (PRA)

A systematic method for assessing three questions that the NRC uses to define "risk." These questions consider (1) what can go wrong, (2) how likely it is, and (3) what its consequences might be. These questions allow the NRC to understand likely outcomes, sensitivities, areas of importance, system interactions, and areas of uncertainty, which the staff can use to identify risk-significant scenarios. The NRC uses PRA to determine a numeric estimate of risk to provide insights into the strengths and weaknesses of the design and operation of a nuclear power plant.

Production expense

Production expense is one component of the cost of generating electric power, which includes costs associated with fuel, as well as plant operation and maintenance.

Rad (radiation absorbed dose)

One of the two units used to measure the amount of radiation absorbed by an object or person, known as the "absorbed dose," which reflects the amount of energy that radioactive sources deposit in materials through which they pass. The radiation-absorbed dose (rad) is the amount of energy (from any type of ionizing radiation) deposited in any medium (e.g., water, tissue, air). An absorbed dose of 1 rad means that 1 gram of material absorbed 100 ergs of energy (a small but measurable amount) as a result of exposure to radiation. The related international system unit is the gray (Gy), where 1 Gy is equivalent to 100 rad.

Radiation, ionizing

A form of radiation, which includes alpha particles, beta particles, gamma rays and x-rays, neutrons, high-speed electrons, and high-speed protons. Compared to nonionizing radiation, such as found in ultraviolet light or microwaves, ionizing radiation is considerably more energetic. When ionizing radiation passes through material such as air, water, or living tissue, it deposits enough energy to break molecular bonds and displace (or remove) electrons. This electron displacement may lead to changes in living cells. Given this ability, ionizing radiation has a number of beneficial uses, including treating cancer or sterilizing medical equipment. However, ionizing radiation is potentially harmful if not used correctly, and high doses may result in severe skin or tissue damage. It is for this reason that the NRC strictly regulates commercial and institutional uses of the various types of ionizing radiation.

Radiation, nuclear

Energy given off by matter in the form of tiny fast-moving particles (alpha particles, beta particles, and neutrons) or pulsating electromagnetic rays or waves (gamma rays) emitted from the nuclei of unstable radioactive atoms. All matter is composed of atoms, which are made up of various parts; the nucleus contains minute particles called protons and neutrons, and the atom's outer shell contains other particles called electrons. The nucleus carries a positive electrical charge, while the electrons carry a negative electrical charge. These forces work toward a strong, stable balance by getting rid of excess atomic energy (radioactivity). In that process, unstable radioactive nuclei may emit energy, and this spontaneous emission is called nuclear radiation.

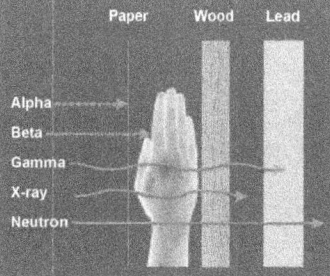

All types of nuclear radiation are also ionizing radiation, but the reverse is not necessarily true; for example, x-rays are a type of ionizing radiation, but they are not nuclear radiation because they do not originate from atomic nuclei. In addition, some elements are naturally radioactive, as their nuclei emit nuclear radiation as a result of radioactive decay, but others become radioactive by being irradiated in a reactor. Naturally occurring nuclear radiation is indistinguishable from induced radiation.

Radiation source

A radioactive material or byproduct that is specifically manufactured or obtained for the purpose of using the emitted radiation. Such sources are commonly used in teletherapy or industrial radiography; in various types of industrial gauges, irradiators, and gamma knives; and as power sources for batteries (such as those used in spacecraft). These sources usually consist of a known quantity of radioactive material, which is encased in a manmade capsule, sealed between layers of nonradioactive material, or firmly bonded to a nonradioactive substrate to prevent radiation leakage. Other radiation sources include devices such as accelerators and x-ray generators.

Radiation standards

Exposure limits; permissible concentrations; rules for safe handling; and regulations regarding receipt, possession, use, transportation, storage, disposal, and industrial control of radioactive material.

Radiation therapy (radiotherapy)

The therapeutic use of ionizing radiation to treat disease in patients. Although most radiotherapy procedures are intended to kill cancerous tissue or reduce the size of a tumor, therapeutic doses may also be used to reduce pain or treat benign conditions. For example, intervascular brachytherapy uses radiation to treat clogged blood vessels. Other common radiotherapy procedures include gamma stereotactic radiosurgery (gamma knife), teletherapy, and iodine treatment to correct an overactive thyroid gland. These procedures use radiation sources, regulated by the NRC and its Agreement States, that may be applied either inside or outside the body. In either case, the goal of radiotherapy is to deliver the required therapeutic or pain-relieving dose of radiation with high precision and for the required length of time, while preserving the surrounding healthy tissue.

Radiation warning symbol

An officially prescribed magenta or black trefoil on a yellow background, which must be displayed where certain quantities of radioactive materials are present or where certain doses of radiation could be received.

Radioactive contamination

Undesirable radioactive material (with a potentially harmful effect) that is either airborne or deposited in (or on the surface of) structures, objects, soil, water, or living organisms (people, animals, or plants) in a concentration that may harm people, equipment, or the environment.

Radioactive decay

The spontaneous transformation of one radioisotope into one or more different isotopes (known as "decay products" or "daughter products"), accompanied by a decrease in radioactivity (compared to the parent material). This transformation takes place over a defined period of time (known as a "half-life"), as a result of electron capture; fission; or the emission of alpha particles, beta particles, or photons (gamma radiation or x-rays) from the nucleus of an unstable atom. Each isotope in the sequence (known as a "decay chain") decays to the next until it forms a stable, less energetic end product. In addition, radioactive decay may refer to gamma-ray and conversion electron emission, which only reduces the excitation energy of the nucleus.

Radioactivity

The property possessed by some elements (such as uranium) of spontaneously emitting energy in the form of radiation as a result of the decay (or disintegration) of an unstable atom. Radioactivity is also the term used to describe the rate at which radioactive material emits radiation. Radioactivity is measured in units of becquerels or disintegrations per second.

Radiography

The use of sealed sources of ionizing radiation for nondestructive examination of the structure of materials. When the radiation penetrates the material, it produces a shadow image by blackening a sheet of photographic film that has been placed behind the material, and the differences in blackening suggest flaws and unevenness in the material.

Radioisotope (radionuclide)

An unstable isotope of an element that decays or disintegrates spontaneously, thereby emitting radiation. Approximately 5,000 natural and artificial radioisotopes have been identified.

Radiopharmaceutical

A pharmaceutical drug that emits radiation and is used in diagnostic or therapeutic medical procedures. Radioisotopes that have short half-lives are generally preferred to minimize the radiation dose to the patient and the risk of prolonged exposure. In most cases, these short-lived radioisotopes decay to stable elements within minutes, hours, or days, allowing patients to be released from the hospital in a relatively short time.

Reactor core

The central portion of a nuclear reactor, which contains the fuel assemblies, water, and control mechanisms, as well as the supporting structure. The reactor core is where fission takes place.

Reactor Oversight Process (ROP)

The process by which the NRC monitors and evaluates the performance of commercial nuclear power plants. Designed to focus on those plant activities that are most important to safety, the process uses inspection findings and performance indicators to assess each plant's safety performance.

Regulation

The governmental function of controlling or directing economic entities through the process of rulemaking and adjudication.

GLOSSARY

Regulatory Information Conference

An annual NRC conference that brings together NRC staff, regulated utilities, materials users, and other interested stakeholders to discuss nuclear safety topics and significant and timely regulatory activities through informal dialogue to ensure an open regulatory process.

Rem (roentgen equivalent man)

One of the two standard units used to measure the dose equivalent (or effective dose), which combines the amount of energy (from any type of ionizing radiation) that is deposited in human tissue with the biological effects of the given type of radiation. For beta and gamma radiation, the dose equivalent is the same as the absorbed dose. By contrast, the dose equivalent is larger than the absorbed dose for alpha and neutron radiation, because these types of radiation are more damaging to the human body. Thus, the dose equivalent (in rems) is equal to the absorbed dose (in rads) multiplied by the quality factor of the type of radiation (Title 10 of the *Code of Federal Regulations* (10 CFR) 20.1004, "Units of Radiation Dose"). The related international system unit is the sievert (Sv), where 100 rem is equivalent to 1 Sv.

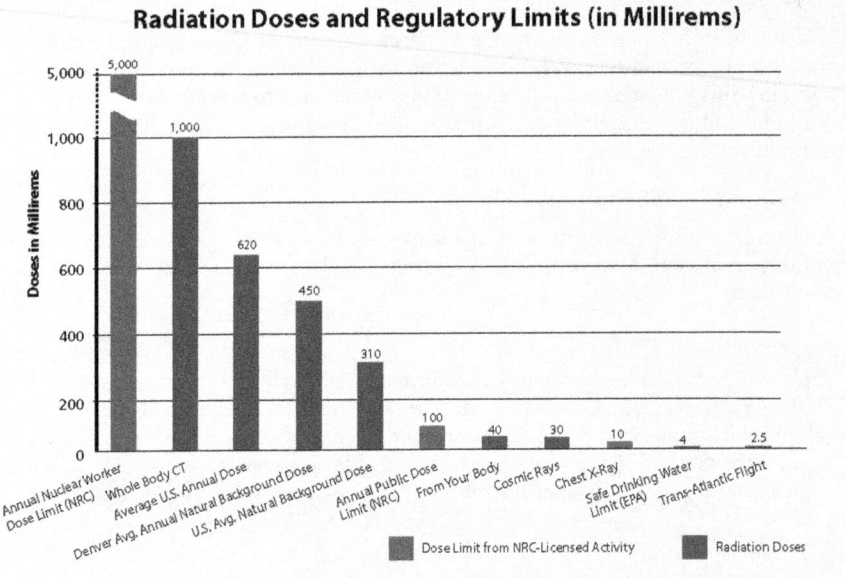

Renewable resources

Natural, but limited, energy resources that can be replenished, including biomass, hydro, geothermal, solar, and wind. These resources are virtually inexhaustible but limited in the amount of energy that is available per unit of time. In the future, renewable resources could also include the use of ocean thermal, wave, and tidal action technologies. Utility renewable resource applications include bulk electricity generation, onsite electricity generation, distributed electricity generation, nongrid-connected generation, and demand-reduction (energy efficiency) technologies. The Information Digest has included conventional hydroelectric and storage hydroelectric in a separate category from other resources.

Risk

The combined answer to three questions that consider (1) what can go wrong, (2) how likely it is, and (3) what its consequences might be. These three questions allow the NRC to understand likely outcomes, sensitivities, areas of importance, system interactions, and areas of uncertainty, which can be used to identify risk-significant scenarios.

Risk-based decisionmaking

An approach to regulatory decisionmaking that considers only the results of a probabilistic risk assessment.

Risk-informed decisionmaking

An approach to regulatory decisionmaking, in which insights from probabilistic risk assessment are considered with other engineering insights.

Risk-informed regulation

An approach to regulation taken by the NRC, which incorporates an assessment of safety significance or relative risk. This approach ensures that the regulatory burden imposed by an individual regulation or process is appropriate to its importance in protecting the health and safety of the public and the environment.

Risk-significant

"Risk-significant" can refer to a facility's system, structure, component, or accident sequence that exceeds a predetermined limit for contributing to the risk associated with the facility. The term also describes a level of risk exceeding a predetermined "significance" level.

Safeguards

The use of material control and accounting programs to verify that all special nuclear material is properly controlled and accounted for, as well as the physical protection (or physical security) equipment and security forces. As used by IAEA, this term also means verifying that the peaceful use commitments made in binding nonproliferation agreements, both bilateral and multilateral, are honored.

Safeguards Information

A special category of sensitive unclassified information that must be protected. Safeguards Information concerns the physical protection of operating power reactors, spent fuel shipments, strategic special nuclear material, or other radioactive material.

Safety-related

In the regulatory arena, this term applies to systems, structures, components, procedures, and controls (of a facility or process) that are relied upon to remain functional during and following design-basis events. Their functionality ensures that key regulatory criteria, such as levels of radioactivity released, are met. Examples of safety-related functions include shutting down a nuclear reactor and maintaining it in a safe-shutdown condition.

GLOSSARY

Safety-significant

When used to qualify an object, such as a system, structure, component, or accident sequence, this term identifies that object as having an impact on safety, whether determined through risk analysis or other means, that exceeds a predetermined significance criterion.

SAFSTOR

A method of decommissioning in which a nuclear facility is placed and maintained in a condition that allows the facility to be safely stored and subsequently decontaminated (deferred decontamination) to levels that permit release for unrestricted use.

Scram

The sudden shutting down of a nuclear reactor, usually by rapid insertion of control rods, either automatically or manually by the reactor operator. Also known as a "reactor trip."

Sensitive unclassified nonsafeguards information

Information that is generally not publicly available and that encompasses a wide variety of categories, such as proprietary information, personal and private information, or information subject to attorney-client privilege.

Shutdown

A decrease in the rate of fission (and heat/energy production) in a reactor (usually by the insertion of control rods into the core).

Source material

Uranium or thorium, or any combination thereof, in any physical or chemical form, or ores that contain, by weight, one-twentieth of one percent (0.05 percent) or more of (1) uranium, (2) thorium, or (3) any combination thereof. Source material does not include special nuclear material.

Special nuclear material

Plutonium, uranium-233, or uranium enriched in the isotopes uranium-233 or uranium-235.

Spent fuel pool

An underwater storage and cooling facility for spent (depleted) fuel assemblies that have been removed from a reactor.

Spent (depleted or used) nuclear fuel

Nuclear reactor fuel that has been used to the extent that it can no longer effectively sustain a chain reaction.

Subcriticality

The condition of a nuclear reactor system, in which nuclear fuel no longer sustains a fission chain reaction (that is, the reaction fails to initiate its own repetition, as it would in a reactor's normal operating condition). A reactor becomes subcritical when its fission events fail to release a sufficient number of neutrons to sustain an ongoing series of reactions, possibly as a result of increased neutron leakage or poisons.

Teletherapy

Treatment in which the source of the therapeutic radiation is at a distance from the body. Because teletherapy is often used to treat malignant tumors deep within the body by bombarding them with a high-energy beam of gamma rays (from a radioisotope such as cobalt-60) projected from outside the body, it is often called "external beam radiotherapy."

Title 10 of the *Code of Federal Regulations* (10 CFR)

Four volumes of the *Code of Federal Regulations* (CFR) address energy-related topics. Parts 1 to 199 contain the regulations (or rules) established by the NRC. These regulations govern the transportation and storage of nuclear materials; use of radioactive materials at nuclear power plants, research and test reactors, uranium recovery facilities, fuel cycle facilities, waste repositories, and other nuclear facilities; and use of nuclear materials for medical, industrial, and academic purposes.

Transient

A change in the reactor coolant system temperature, pressure, or both, attributed to a change in the reactor's power output. Transients can be caused by (1) adding or removing neutron poisons, (2) increasing or decreasing electrical load on the turbine generator, or (3) accident conditions.

Transuranic waste

Material contaminated with transuranic elements—artificially made, radioactive elements, such as neptunium, plutonium, americium, and others—that have atomic numbers higher than uranium in the periodic table of elements. Transuranic waste is primarily produced from recycling spent fuel or using plutonium to fabricate nuclear weapons.

Tritium

A radioactive isotope of hydrogen. Because it is chemically identical to natural hydrogen, tritium can easily be taken into the body by any ingestion path. It decays by emitting beta particles and has a half-life of about 12.5 years.

Uprate

See *Power uprate*.

Uranium

A radioactive element with the atomic number 92 and, as found in natural ores, an atomic weight of approximately 238. The two principal natural isotopes are uranium-235 (which comprises 0.7 percent of natural uranium), which is fissile, and uranium-238 (99.3 percent of natural uranium), which is fissionable by fast neutrons and is fertile, meaning that it becomes fissile after absorbing one neutron. Natural uranium also includes a minute amount of uranium-234.

Uranium fuel fabrication facility

A facility that converts enriched UF_6 into fuel for commercial light-water power reactors, research and test reactors, and other nuclear reactors. The UF_6, in solid form in containers, is heated to a gaseous form and then chemically processed to form uranium dioxide (UO_2) powder. This powder is then processed into ceramic pellets and loaded into metal tubes, which are subsequently bundled into fuel assemblies. Fabrication also can involve MOX fuel, which contains plutonium oxide mixed with either natural or depleted uranium oxide, in ceramic pellet form.

Uranium hexafluoride production facility (or uranium conversion facility)

A facility that receives natural uranium in the form of ore concentrate (known as "yellowcake") and converts it into UF_6 in preparation for fabricating fuel for nuclear reactors.

U.S. Department of Energy (DOE)

The Federal agency established by Congress to advance the national, economic, and energy security of the United States, among other missions.

U.S. Department of Homeland Security (DHS)

The Federal agency responsible for leading the unified national effort to secure the United States against those who seek to disrupt the American way of life. DHS is also responsible for preparing for and responding to all hazards and disasters and includes the formerly separate FEMA, the Coast Guard, and the Secret Service.

U.S. Environmental Protection Agency (EPA)

The Federal agency responsible for protecting human health and safeguarding the environment. The EPA leads the Nation's environmental science, research, education, and assessment efforts to ensure that efforts to reduce environmental risk are based on the best available scientific information. The EPA also ensures that environmental protection is an integral consideration in U.S. policies.

Viability assessment

A decisionmaking process used by DOE to assess the prospects for safe and secure permanent disposal of HLW in an excavated, underground facility, known as a geologic repository. This decisionmaking process is based on (1) specific design work on the critical elements of the repository and waste package, (2) a total system performance assessment that will describe the probable behavior of the repository, (3) a plan and cost estimate for the work required to complete the license application, and (4) an estimate of the costs to construct and operate the repository.

Waste, radioactive

Radioactive materials at the end of their useful life or in a product that is no longer useful and requires proper disposal.

Waste classification (classes of waste)

Classification of LLW according to its radiological hazard. The classes include Class A, B, and C, with Class A being the least hazardous and accounting for 96 percent of LLW. As the waste class and hazard increase, the regulations established by the NRC require progressively greater controls to protect the health and safety of the public and the environment.

Watt

A unit of power (in the international system of units) defined as the consumption or conversion of one joule of energy per second. In electricity, a watt is equal to current (in amperes) multiplied by voltage (in volts).

Watthour

An unit of energy equal to one watt of power steadily supplied to, or taken from, an electrical circuit for one hour (or exactly 3.6×10^3 joules).

Well-logging

All operations involving the lowering and raising of measuring devices or tools that contain licensed nuclear material or are used to detect licensed nuclear materials in wells for the purpose of obtaining information about the well or adjacent formations that may be used in oil, gas, mineral, ground water, or geological exploration.

Wheeling service

The movement of electricity from one system to another over transmission facilities of intervening systems. Wheeling service contracts can be established between two or more systems.

Yellowcake

The solid form of mixed uranium oxide, which is produced from uranium ore in the uranium recovery (milling) process. The material is a mixture of uranium oxides, which can vary in proportion and color from yellow to orange to dark green (blackish) depending on the temperature at which the material is dried (which affects the level of hydration and impurities), with higher drying temperatures producing a darker and less soluble material. (The yellowcake produced by most modern mills is actually brown or black, rather than yellow, but the name comes from the color and texture of the concentrates produced by early milling operations.) Yellowcake is commonly referred to as U_3O_8, because that chemical compound comprises approximately 85 percent of the yellowcake produced by uranium recovery facilities, and that product is then transported to a uranium conversion facility, where it is transformed into UF_6, in preparation for fabricating fuel for nuclear reactors.

GLOSSARY

WEB LINK INDEX

NRC: An Independent Regulatory Agency

Mission, Goals, and Statutory Authority

Strategic Plan Fiscal Years 2008–2013
www.nrc.gov/reading-rm/doc-collections/nuregs/staff/sr1614/v4/sr1614v4.pdf

Statutory Authority
www.nrc.gov/about-nrc/governing-laws.html

Major Activities

Public Involvement
www.nrc.gov/public-involve.html

Freedom of Information Act (FOIA)
www.nrc.gov/reading-rm/foia/foia-privacy.html

Agency Rulemaking Actions
www.regulations.gov

Significant Enforcement Actions
www.nrc.gov/reading-rm/doc-collections/enforcement/actions/

Organizations and Functions

Organization Chart
www.nrc.gov/about-nrc/organization/nrcorg.pdf

The Commission
www.nrc.gov/about-nrc/organization/commfuncdesc.html

Commission Direction-Setting and Policymaking Activities
www.nrc.gov/about-nrc/policymaking.html

NRC Regions
www.nrc.gov/about-nrc/locations.html

NRC Budget

Performance Budget: Fiscal Year 2009 (NUREG-1100, Vol. 24)
www.nrc.gov/reading-rm/doc-collections/nuregs/staff/sr1100/v24/

U.S. and Worldwide Nuclear Energy

U.S. Electricity

Energy Information Administration
Official Energy Statistics from the U.S. Government
www.eia.doe.gov

Worldwide Electricity Generated by Commercial Nuclear Power

International Atomic Energy Agency (IAEA)
www.iaea.org

IAEA Power Reactor Information System (PRIS)
www.iaea.org/programmes/a2

Nuclear Energy Agency (NEA)
www.nea.fr/

World Nuclear Association (WNA)
www.world-nuclear.org/

World Nuclear Power Reactors and Uranium Requirements
www.world-nuclear.org/info/reactors.html

WNA Reactor Database
www.world-nuclear.org/reference/default.aspx

WNA Global Nuclear Reactors Map
www.wano.org.uk/WANO_Documents/WANO_Map/WANO_Map.pdf

NRC Office of International Programs
www.nrc.gov/about-nrc/organization/oipfuncdesc.html

NRC 20th Regulatory Information Conference (RIC)
www.nrcric.org

International Activities

Treaties and Conventions
www.nrc.gov/about-nrc/ip/treaties-conventions.html

Operating Nuclear Reactors

U.S. Commercial Nuclear Power Reactors

Commercial Reactors
www.nrc.gov/info-finder/reactor/

Oversight of U.S. Commercial Nuclear Power Reactors

Reactor Oversight Process (ROP)
www.nrc.gov/NRR/OVERSIGHT/ASSESS/index.html

NUREG-1649, "Reactor Oversight Process"
www.nrc.gov/reading-rm/doc-collections/nuregs/staff/sr1649/r4/

ROP Performance Indicators Summary
www.nrc.gov/NRR/OVERSIGHT/ASSESS/pi_summary.html

Future U.S. Commercial Nuclear Power Reactor Licensing

New Reactor License Process
www.nrc.gov/reactors/new-reactor-op-lic/licensing-process.html#licensing

New Reactors

New Reactor Licensing
www.nrc.gov/reactors/new-reactors.html

Reactor License Renewal

Reactor License Renewal Process
www.nrc.gov/reactors/operating/licensing/renewal/process.html

WEB LINK INDEX

10 CFR Part 51, "Environmental Protection Regulations for Domestic Licensing and Related Regulatory Functions"
www.nrc.gov/reading-rm/doc-collections/cfr/part051/

10 CFR Part 54, "Requirements for Renewal of Operating Licenses for Nuclear Power Plants"
www.nrc.gov/reading-rm/doc-collections/cfr/part054/

Status of License Renewal Applications and Industry Activities
www.nrc.gov/reactors/operating/licensing/renewal/applications.html

U.S. Nuclear Research and Test Reactors

Research and Test Reactors
www.nrc.gov/reactors/non-power.html

Nuclear Regulatory Research

Nuclear Reactor Safety Research
www.nrc.gov/about-nrc/regulatory/research/reactor-rsch.html

State-of-the-Art Reactor Consequence Analyses (SOARCA)
www.nrc.gov/about-nrc/regulatory/research/soar.html

Risk Assessment in Regulation
www.nrc.gov/about-nrc/regulatory/risk-informed.html

Digital Instrumentation and Controls
www.nrc.gov/about-nrc/regulatory/research/digital.html

Computer Codes
www.nrc.gov/about-nrc/regulatory/research/comp-codes.html

Generic Issues Program
www.nrc.gov/about-nrc/regulatory/gen-issues.html

The Committee To Review Generic Requirements (CRGR)
www.nrc.gov/about-nrc/regulatory/crgr.html

Nuclear Materials

U.S. Fuel Cycle Facilities

U.S. Fuel Cycle Facilities
www.nrc.gov/info-finder/materials/fuel-cycle/

Uranium Recovery

Uranium Milling/Recovery
www.nrc.gov/info-finder/materials/uranium/

U.S. Materials Licenses

Materials Licensees Toolkits
www.nrc.gov/materials/miau/mat-toolkits.html

WEB LINK INDEX

Information Security

Information Security
www.nrc.gov/security/info-security.html

Assuring the Security of Radioactive Material
www.nrc.gov/security/byproduct.html

Emergency Preparedness and Response

Emergency Preparedness and Response
www.nrc.gov/about-nrc/emerg-preparedness.html

Research and Test Reactor Emergency Preparedness

Research and Test Reactors
www.nrc.gov/reactors/non-power.html

Stakeholder Meetings and Workshops

www.nrc.gov/public-involve/public-meetings/stakeholder-mtngs-wksps.html

Emergency Action Level Development

www.nrc.gov/about-nrc/emerg-preparedness/about-emerg-preparedness/emerg-action-level-dev.html

Hostile Action Based Emergency Preparedness Drills

www.nrc.gov/about-nrc/emerg-preparedness/respond-to-emerg/hostile-action.html

Exercise Schedules

NRC Participation Exercise Schedule
www.nrc.gov/about-nrc/emerg-preparedness/about-emerg-preparedness/exercise-schedules.html

Other Web Links

Employment Opportunities

NRC—*A Great Place to Work*
www.nrc.gov/about-nrc/employment.html

Glossary

NRC Basic References
www.nrc.gov/reading-rm/basic-ref/glossary/full-text.html

Glossary of Electricity Terms

www.eia.doe.gov/cneaf/electricity/epav1/glossary.html

Glossary of Security Terms

https://hseep.dhs.gov/DHSResource/Glossary.aspx

Public Involvement

NRC Library
www.nrc.gov/reading-rm.html

Freedom of Information and Privacy Acts
www.nrc.gov/reading-rm/foia/foia-privacy.html

Agencywide Documents Access and Management System (ADAMS)
www.nrc.gov/reading-rm/adams.html

Public Document Room
www.nrc.gov/reading-rm/pdr.html

Public Meeting Schedule
www.nrc.gov/public-involve/public-meetings/index.cfm

Documents for Comments
www.nrc.gov/public-involve/doc-comment.html

Small Business and Civil Rights

Contracting Opportunities for Small Businesses
www.nrc.gov/about-nrc/contracting/small-business.html

Workplace Diversity
www.nrc.gov/about-nrc/employment/diversity.html

Discrimination Complaint Activity
www.nrc.gov/about-nrc/civil-rights.html

Equal Employment Opportunity Program
www.nrc.gov/about-nrc/civil-rights/eeo.html

Limited English Proficiency
www.nrc.gov/about-nrc/civil-rights/limited-english.html

Minority Serving Institutions Program
www.nrc.gov/about-nrc/grants.html#msip

NRC Comprehensive Diversity Management Plan brochure
www.nrc.gov/reading-rm/doc-collections/nuregs/brochures/br0316

NRC Mentoring Program
www.nrc.gov/about-nrc/employment/diversity.html/#mentor

INDEX

A

Academic v, 2, 3, 8, 55, 58, 59, 61, 158, 169, 182

Appropriation 11

Agreement States 4, 58 (3)

Atomic Energy Act 3, 8, 22, 47, 82, 146, 181

B

Bilateral programs 22

Boiling-water reactor vii, 28, 31, 44, 98, 99, 145, 149, 159, 162, 182

Byproduct 2, 116, 145, 148, 155, 156, 159, 160, 164, 176, 181

C

Capacity v, vi, viii, 14, 15, 17, 18, 19, 77, 134, 135, 142, 146, 154, 157, 182,

Cask 73, 76, 77, 79, 80, 83, 129, 146, 148, 149

D

Decommissioning 3, 4, 8, 12, 34, 51, 52, 66, 67, 68, 84, 85, 115, 116, 127, 133, 140, 147, 148, 150, 168, 175, 181, 182, 183, v, vi, vii

Disposal v, vii, 2, 4, 8, 64, 65, 68, 74, 75, 79, 82, 146, 148, 152, 154, 156, 158, 164, 170, 175, 182

E

Energy Reorganization Act 2, 3, 181,

Enrichment vii, 4, 58, 65, 68, 69, 70, 99, 152, 153, 160, 182

Environmental issues 50,

F

Fabrication 58, 65, 68, 69, 71, 88, 145, 152, 157, 160, 169, 170, 182, vii,

Fuel facilities 9, 58, 68, 88,

Fuel rods 37, 71, 78, 152, 153, 162

G

Gaseous diffusion 69, 70, 153, 182,

Gauges 61, 62, 153, 164, 175

Gross generation 158

H

Hexafluoride 68, 69, 152, 153, 169, 170, 171, 182

High-level radioactive waste v, 4, 12, 76, 79, 82, 146, 154, 155, 170, 175, 182

I

Industrial v, 2, 3, 8, 25, 26, 58, 61, 150, 151, 158, 160, 164, 169, 175, 182

In situ leach 155

Inspection vii, 3, 8, 9, 23, 27, 33, 34, 44, 45, 46, 50, 52, 65, 68, 71, 72, 83, 88, 89, 138, 139, 140, 151, 158, 161, 162, 165, 182, 183

International iii, v, 7, 13, 20, 21, 22, 24, 25, 55, 69, 89, 98, 99, 129, 130, 131, 132, 133, 137, 143, 155, 157, 163, 166, 167, 171, 172, 173, 180, 181

L

Licensing iii, v, vii, 3, 4, 6, 8, 9, 22, 23, 24, 28, 33, 39, 41, 42, 43, 45, 46, 48, 50, 52, 65, 68, 72, 77, 82, 83, 124, 140, 143, 156, 173, 174, 180, 181, 182, 183

Low-level radioactive waste v, viii, 4, 8, 74, 75, 147, 156, 170, 175, 181, 182

M

Materials v, vii, viii, 2, 3, 4, 5, 7, 8, 9, 11, 12, 22, 24, 51, 52, 54, 55, 57, 58, 59, 60, 61, 62, 63, 64, 65, 72, 83, 84, 88, 89, 94, 95, 138, 139, 140, 145, 146, 147, 148, 151, 155, 156, 158, 159, 160, 163, 164, 165, 166, 169, 170, 171, 174, 175, 181, 182, 183, 184

Medical v, vii , 2, 3, 8, 25, 58, 59, 60, 61, 63, 74, 138, 139, 140, 145, 150, 155, 156, 158, 160, 163, 165, 169, 175, 182

NRC FORM 335 (2-89) NRCM 1102, 3201,3202	U.S. NUCLEAR REGULATORY COMMISSION **BIBLIOGRAPHIC DATA SHEET** *(See instructions on the reverse)*	1. REPORT NUMBER (Assigned by NRC, Add Vol., Supp., Rev., and Addendum Numbers, if any.) NUREG-1350, Vol. 23

2. TITLE AND SUBTITLE U.S. Nuclear Regulatory Commission Information Digest 2011-2012	3. DATE REPORT PUBLISHED	
	MONTH	YEAR
	August	2011
	4. FIN OR GRANT NUMBER n/a	

5. AUTHOR(S) Ivonne Couret, et al.	6. TYPE OF REPORT Annual
	7. PERIOD COVERED *(Inclusive Dates)* 2010-2011

8. PERFORMING ORGANIZATION - NAME AND ADDRESS *(If NRC provide Division, Office or Region, U.S. Nuclear Regulatory Commission, and mailing address; if contractor provide name and mailing address.)*

Public Affairs Staff
Office of Public Affairs
U.S. Nuclear Regulatory Commission
Washington, DC 20555-0001

9. SPONSORING ORGANIZATION - NAME AND ADDRESS *(If NRC, type "Same as above"; if contractor, provide NRC Division, Office or Region, U.S. Nuclear Regulatory Commission, and mailing address.)*

Same as 8, above

10. SUPPLEMENTARY NOTES
There may be a supplementary document produced with only the updated figures, tables, and/or appendices within the next 12 months.

11. ABSTRACT *(200 words or less)*

The U.S. Nuclear Regulatory Commission (NRC) 2011–2012 Information Digest provides a summary of information about the NRC and the industry it regulates. It describes the agency's regulatory responsibilities and licensing activities and also provides general information on nuclear-related topics. It is updated annually.

The Information Digest includes NRC- and industry-related data in a quick reference format. Data include activities through 2010 or the most current data available at manuscript completion. The Web Link Index provides URL addresses for more information on major topics. The Digest also includes a tear-out reference sheet, the NRC Facts at a Glance.

The NRC reviewed information from industry and international sources but did not perform an independent verification. This edition contains adjustments to preliminary figures from the previous year. All information is final unless otherwise noted.

The NRC is the source for all photographs, graphics, and tables unless otherwise noted.

The agency welcomes comments or suggestions on the Information Digest. Please contact Ivonne Couret by mail at the Office of Public Affairs, U.S. Nuclear Regulatory Commission, Washington, DC 20555-0001 or by e-mail at OPA.Resource@nrc.gov.

12. KEY WORDS/DESCRIPTORS *(List words or phrases that will assist researchers in locating the report.)* Information Digest 2011-2012 Edition NRC Facts Nuclear Regulatory Commission	13. AVAILABILITY STATEMENT unlimited
	14. SECURITY CLASSIFICATION
	(This Page) unclassified
	(This Report) unclassified
	15. NUMBER OF PAGES 196
	16. PRICE

NRC FORM 335 (9-2004)   PRINTED ON RECYCLED PAPER

MISSION

The mission of the U.S. Nuclear Regulatory Commission (NRC) is to license and regulate the Nation's civilian use of byproduct, source, and special nuclear materials to ensure adequate protection of public health and safety, to promote the common defense and security, and to protect the environment.

COMMISSION

Chairman Gregory B. Jaczko
  Term Ends June 30, 2013
Commissioner Kristine L. Svinicki
  Term Ends June 30, 2012
Commissioner George Apostolakis
  Term Ends June 30, 2014
Commissioner William D. Magwood, IV
  Term Ends June 30, 2015
Commissioner William C. Ostendorff
  Term Ends June 30, 2011

LOCATIONS

Headquarters:
U.S. Nuclear Regulatory Commission
Rockville, MD
301-415-7000 1-800-368-5642
One White Flint North: 11555 Rockville Pike
Two White Flint North: 11545 Rockville Pike
Executive Boulevard Building: 6003 Executive Boulevard
Gateway Building: 7201 Wisconsin Ave
Twinbrook Building: 12300 Twinbrook Parkway
Church Street Building: 21 Church Street

Operations Center:
Rockville, MD
301-816-5100
The NRC maintains an operations center that State agencies and other Federal agencies coordinate with concerning operating events at commercial nuclear facilities. NRC operations officers staff the operations center 24 hours a day.

Regional Offices:

Region I	Region III
King of Prussia, PA	Lisle, IL
610-337-5000	630-829-9500
Region II	Region IV
Atlanta, GA	Arlington, TX
404-997-4000	817-860-8100

High-Level Waste Management Office
Las Vegas, NV
702-794-5048

Training and Professional Development:
Technical Training Center
Chattanooga, TN
423-855-6500

Professional Development Center
Bethesda, MD
301-492-2000

Resident Sites:
At least two NRC resident inspectors who report to the appropriate regional office are located at each nuclear power plant site.

NRC BUDGET

- Total authority: $1,054.1 million
- Total staff: 3,992
- Budget amount expected to be recovered by annual fees to licensees: $915.8 million
- NRC research program support: $61 million

NRC REGULATORY ACTIVITIES

- Regulation and guidance—rulemaking
- Policymaking
- Licensing, decommissioning, and certification
- Research
- Oversight
- Emergency preparedness and response
- Support of Commission decisions

NRC GOVERNING LEGISLATION

The NRC was established by the Energy Reorganization Act of 1974. A summary of laws that govern the agency's operations is provided below. The text of other laws may be found in NUREG-0980, "Nuclear Regulatory Legislation."

FUNDAMENTAL LAWS GOVERNING CIVILIAN USES OF RADIOACTIVE MATERIALS

Nuclear Materials and Facilities

- Atomic Energy Act of 1954, as amended
- Energy Reorganization Act of 1974

Radioactive Waste

- Nuclear Waste Policy Act of 1982, as amended
- Low-Level Radioactive Waste Policy Amendments Act of 1985
- Uranium Mill Tailings Radiation Control Act of 1978

Nonproliferation

- Nuclear Non-Proliferation Act of 1978

FUNDAMENTAL LAWS GOVERNING THE PROCESSES OF REGULATORY AGENCIES

- Administrative Procedure Act (5 U.S.C. Chapters 5 through 8)
- National Environmental Policy Act
- Diplomatic Security and Anti-Terrorism Act of 1986
- Energy Policy Act of 1992
- Energy Policy Act of 2005

TREATIES AND AGREEMENTS

- Nuclear Non-Proliferation Treaty
- International Atomic Energy Agency/U.S. Safeguards Agreement
- Convention on the Physical Protection of Nuclear Material
- Convention on Early Notification of a Nuclear Accident
- Convention on Assistance in Case of a Nuclear Accident and Radiological Emergency
- Convention on Nuclear Safety
- Convention on Supplemental Liability and Safety of Spent Fuel Management and on the Safety of Radioactive Waste Management

U.S. COMMERCIAL NUCLEAR POWER REACTORS

- Generate 20 percent of the Nation's electricity
- Operate in 31 States
- 104 nuclear power plants licensed to operate in the United States
  - 69 pressurized-water reactors
  - 35 boiling-water reactors
- 4 reactor fuel vendors
- 26 parent companies
- 80 different designs
- 65 commercial reactor sites
- 14 decommissioning power reactors
- 6,055 total inspection hours in calendar year 2010 at operating reactors; approximately 3,000 source documents concerning events reviewed

REACTOR LICENSE RENEWAL

Commercial power reactor operating licenses are valid for 40 years and may be renewed for up to an additional 20 years.

- 41 sites and 71 units with renewal licenses issued at operating nuclear plants
- 9 sites with license renewal applications in review
- 12 sites with letters of intent to submit renewal license applications

NEW REACTOR LICENSE PROCESS

Early Site Permit (ESP)

- 4 ESPs issued
- 2 ESP applications in review

Combined License—Construction and Operating (COL)

- 18 COL applications received/docketed for 28 units; of these, 13 applications are under active review

Reactor Design Certification (DC)

- 4 DCs issued
- 5 DCs in review

NUCLEAR RESEARCH AND TEST REACTORS

- 43 licensed research reactors and test reactors
  - 31 reactors operating in 22 States
  - 12 reactors permanently shut down and in various stages of decommissioning (since 1958, a total of 82 licensed research and test reactors have been decommissioned)

NUCLEAR SECURITY AND SAFEGUARDS

- Once every 2 years, each nuclear power plant performs full-scale emergency preparedness exercises.
- Plants also conduct additional emergency drills between full-scale exercises. The NRC evaluates emergency exercises and drills.

NUCLEAR MATERIALS

- The NRC and the Agreement States issue approximately 22,000 licenses for medical, academic, industrial, and general uses of nuclear materials.
- The NRC administers approximately 3,000 licenses.
- 37 Agreement States administer approximately 19,000 licenses.

16 Uranium Recovery Sites Licensed by the NRC

- 5 in situ recovery
- 11 conventional recovery (10 are undergoing decommissioning)

15 Fuel Cycle Facilities

- 1 uranium hexafluoride production facility
- 6 uranium fuel fabrication facilities
- 2 gaseous diffusion uranium enrichment facilities (1 in cold shutdown and in process of decertification)
- 3 gas centrifuge uranium enrichment facilities (1 operating with further construction, 1 under construction, and 1 under review)
- 1 mixed-oxide fuel fabrication facility (under construction and review)
- 1 laser separation enrichment facility (under review)
- 1 uranium hexaflouride deconversion facility (under review)
- 180 NRC-licensed facilities authorized to possess plutonium and enriched uranium with inventory registered in the Nuclear Material Management and Safeguards System database

RADIOACTIVE WASTE

Low-Level Radioactive Waste

- 10 regional compacts
- 3 active licensed disposal facilities, 1 expected to receive low-level waste in 2011
- 4 closed disposal facilities

HIGH-LEVEL RADIOACTIVE WASTE MANAGEMENT

Disposal and Storage

- On January 29, 2010, the President created the Blue Ribbon Commission on America's Nuclear Future to reassess the national policy on high-level waste disposal. The task of the Blue Ribbon Commission is to "conduct a comprehensive review of policies for managing the back end of the nuclear fuel cycle." In light of these developments, the NRC began reassessing its management of spent fuel regulations to position the agency to quickly adapt to changes in national policy. The three key areas in this effort are the nuclear fuel cycle, spent fuel storage and transportation, and high-level waste disposal. The NRC is currently conducting orderly closeout of the Yucca Mountain project to be completed at the end of this year.

- The Nuclear Waste Policy Act of 1982, as amended, defines the roles of the three Federal agencies responsible for nuclear waste. The U.S. Department of Energy is responsible for developing permanent disposal capacity for spent fuel and other high-level radioactive waste. The U.S. Environmental Protection Agency (EPA) is responsible for developing environmental standards to evaluate the safety of a geologic repository. The NRC is responsible for developing regulations to implement the EPA safety standards and for licensing the repository.

Spent Nuclear Fuel Storage

- 63 licensed/operating independent spent fuel storage installations in 33 States
- 15 site-specific licenses
- 48 general licenses

Transportation—Principal Licensing and Inspection Activities

- The NRC examines transport-related safety during approximately 1,000 safety inspections of fuel, reactor, and materials licensees annually.
- The NRC reviews, evaluates, and certifies approximately 80 new, renewal, or amended container-design applications for the transport of nuclear materials annually.
- The NRC reviews and evaluates approximately 150 license applications for the import/export of nuclear materials from the United States annually.
- The NRC inspects about 20 dry storage and transport package licensees annually.

Decommissioning

Approximately 200 material licenses are terminated each year. The NRC's decommissioning program focuses on the termination of licenses that are not routine and that require complex activities.

- 29 nuclear power reactors permanently shut down or in the decommissioning process

- 12 research and test reactors
- 20 complex decommissioning materials facilities
- 1 fuel cycle facility (partial decommissioning)
- 11 uranium recovery facilities in safe storage under NRC jurisdiction

PUBLIC MEETINGS AND INVOLVEMENT

- The NRC conducts 900 public meetings annually.
- The NRC hosts both the Regulatory Information Conference and the Fuel Cycle Information Exchange annually, where participants discuss the latest technical issues.
- The Advisory Committee on Reactor Safeguards held 10 full committee meetings and approximately 70 subcommittee meetings.

NEWS AND INFORMATION

- NRC news releases are available through a free listserv subscription at www.nrc.gov/public-involve/listserver.html.
- Agency photos and videos are available at www.nrc.gov/reading-rm/photo-gallery.

NRC Regions

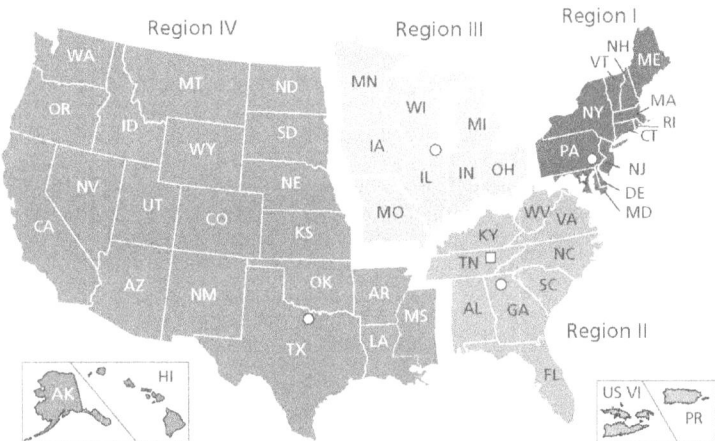

☆ Headquarters (1)
○ Regional Office (4)
□ Technical Training Center (1)

Nuclear Power Plants

- Each regional office oversees the plants in its region except the Grand Gulf plant in Mississippi and the Callaway plant in Missouri, which Region IV oversees.

Material Licensees

- Region I oversees licensees and Federal facilities located geographically in Region I and Region II.
- Region III oversees licensees and Federal facilities located geographically in Region III.
- Region IV oversees licensees and Federal facilities located geographically in Region IV.

Nuclear Fuel Processing Facilities

- Region II oversees all the fuel processing facilities in the region and those in Illinois, New Mexico, and Washington.
- In addition, Region II handles all construction inspectors' activities for new nuclear power plants and fuel cycle facilities in all regions.

CONTACT US

Mailing Address
U.S. Nuclear Regulatory Commission
Washington, DC 20555-0001
1-800-368-5642, 301-415-7000
TTD: 301-415-5575

Delivery Address
11555 Rockville Pike or
11545 Rockville Pike,
Rockville, MD 20852

Public Affairs
301-415-8200
Fax: 301-415-3716

Public Document Room
1-800-397-4209
Fax: 301-415-3548
TDD: 1-800-635-4512

EMPLOYMENT
Human Resources
301-415-7400

Office of the General Counsel Intern Program or Honor Law Graduate Programs
301-415-1515

CONTRACTING OPPORTUNITIES
Small Business Office
1-800-903-7272

FOLLOW US

Blog   Video   Twitter
www.nrc.gov

REPORT A CONCERN

Emergency
Involving a nuclear facility or radioactive materials, including the following:
- Any accident involving a nuclear reactor, nuclear fuel facility, or radioactive materials, or lost or damaged radioactive materials
- Any threat, theft, smuggling, vandalism, or terrorist activity involving a nuclear facility or radioactive materials

**Call the NRC's 24-Hour
Headquarters Operations Center:
301-816-5100**

We accept collect calls.

Note: Calls to this number are recorded.

Nonemergency
Including any concern involving a nuclear reactor, nuclear fuel facility, or radioactive materials.

You may send an e-mail to Allegations.
However, because e-mail transmission may not be completely secure, if you are concerned about protecting your identity it is preferable that you contact us by phone or in person.

You may contact any NRC employee (including a resident inspector) or call:

**NRC's Toll-Free Safety Hotline:
800-695-7403**

Note: Calls to this number are not recorded between the hours of 7:00 a.m. to 5:00 p.m. Eastern Time. However, calls received outside these hours are answered by the Incident Response Operations Center on a recorded line.

Some materials and activities are regulated by Agreement States and States.

**NRC's OIG Hotline:
1-800-233-3497
TDD: 1-800-270-2787
7:00 a.m. - 4:00 p.m. (EST)
After hours, please leave a message.**

The Office of the Inspector General (OIG) at NRC established the Hotline (1-800-233-3497) program to provide the NRC employee, other government employee, licensee/utility employee, contractor employee, and the public with a confidential means of reporting incidences of suspicious activity to OIG concerning fraud, waste, abuse, and employee or management misconduct. Mismanagement of agency programs or danger to public health and safety may also be reported through the Hotline.

It is not our policy to attempt to identify people contacting the Hotline. People may contact OIG by telephone, through an online form, or by mail. There is no caller identification feature associated with the Hotline or any other telephone line in the Inspector General's office. No identifying information is captured when you submit an online form. You may provide your name, address, or phone number, if you wish.

AVAILABILITY OF REFERENCE MATERIALS
IN NRC PUBLICATIONS

U.S. Nuclear Regulatory Commission
NUREG-1350, Volume 23
August 2011